THE ART OF PROBLEM SOLVING IN ORGANIC CHEMISTRY

THE ART OF PROBLEM SOLVING IN ORGANIC CHEMISTRY

Miguel E. Alonso

Department of Chemistry
Institute of Scientific Research of Venezuela
Caracas, Venezuela

A Wiley-Interscience Publication

JOHN WILEY & SONS

New York • Chichester • Brisbane • Toronto • Singapore

To Adela and Gabriel
in this world,
Christiane and Ramon
in the other

Library of Congress Cataloging in Publication Data:

Alonso, Miguel E.
The art of problem solving in organic chemistry.

"A Wiley-Interscience publication."
Includes index.
1. Chemistry, Organic. 2. Problem solving.
I. Title.

QD251.2.A46 1987 547 86-13349
ISBN 0-471-84784-4

Printed in the United States of America

10 9 8 7 6 5 4 3 2

PREFACE

"Science does not prove anything at all; rather it disproves a great deal," asserted K. Popper in *The Logic of Scientific Discovery*. This remarkable thought has triggered a considerable amount of philosophical discussion throughout the world, and its full meaning may be debated for several years. Among other possibilities, this sentence implies that scientific discovery is more solidly developed on the basis of the experimental negation or disapproving of models or working hypotheses that attempt to explain a given phenomenon than on the basis of affirmation by experiment of these models or hypotheses.

The attitude associated with approval is generally recognized as requiring much less effort than that associated with dissent, because the latter implies a more complex thought mechanism that includes analysis, synthesis, selection, comparison, construction of opposing standpoints, and clear verbal composition to express and defend the disagreement. Therefore, Popper's sentence may also be interpreted in terms of a desirable profile for a professional scientist. That is, a person endowed not only with high level cognitive memory or recall thinking, but also with considerable ability for *critical thinking*, which enables him or her to design hypotheses and experiments intended to negate existing models.

The latter quality has been condensed by Howard Schneiderman, Monsanto's vice president for research, in a recent college commencement address (*Chemical and Engineering News*, June 21, 1982), as three essential abilities: development of good taste, ability to communicate in clear language, and a great deal of *problem solving capacity*.

It is clear that the system of scientific education shows inadequacies in at least these three aspects and this lack is currently the cause of deep concern among educators and theoreticians of education. Of these three abilities, problem solving is probably the most important since it should permit the development of analytical skills, synthetic reasoning, discernment in separating the important from the unworthy, and the ability to recognize valid solutions from a variety of alternatives. These qualities help considerably in attaining insight, cleverness, and even artfulness and good taste in professional practice in academic and most industrial environments.

The question then becomes, which mechanism should we adopt to educate students properly in this area and thus overcome this deficiency? There is no unique answer or magic formula. However, a good beginning is the intense practice of problem solving in specific areas of knowledge, although it would be desirable to have a more general syllabus of widespread applicability, at least in the hard core sciences.

And, there is chemistry. In the words of Robertus Alexander Todd, better known as Lord Todd, "there is no question ... that chemistry is the center point of science." I may add that organic chemistry is perhaps the heart of this center point because it underlies so many disciplines, from agricultural production at all levels, biochemistry, industrial chemistry, polymers, pharmaceuticals, to 99% of the chemistry involved in all living systems. Furthermore, the multitude of mechanisms by which organic compounds undergo transformation offers an ideal platform on which those desirable skills mentioned previously can be developed. It is the purpose of this book to construct from this basis the educational means of achieving the development of problem solving skills in the student of advanced organic chemistry. It is also possible that practicing professionals might find this work useful if their exposure to problem solving during their college and university studies has been inadequate.

The use of a number of examples that constitute the series of 56 problems collected and discussed in the third chapter was preferred over long theoretical descriptions. Some necessary fundamental concepts are concentrated in the introductory chapters. This book may be found useful not only as a study guide but also as a source of interesting and somewhat challenging problems and as illustrations of reactions and phenomena of general interest.

I want to express my gratitude to all those who read all or parts of the rough drafts, offering helpful comments. I am particularly thankful to Professor Bruce Ganem and Professor Jerrold Meinwald for their useful suggestions and to Paul Gassman for his advice during the early stages of this work. I especially wish to thank Mrs. Shirley Thomas for her dedicated Production work, Ms. Cheryl Bush for her advice on language usage, and to all my students who, over the years, have provided useful feedback for many of the ideas expressed in this work. Finally, my thanks to the Tarnawiecki family of Lima, Peru. This book benefited greatly from the stimulating and highly caring environment they provided while the writing of the first draft was in progress. Two most unusual people contributed the most to this environment, Don Rafael and my wife, Adela.

MIGUEL E. ALONSO

Caracas, Venezuela
March 1986

CONTENTS

INTRODUCTION

Few persons, if any, will argue convincingly against the premise that problem solving is one of the best means currently available to educate future professionals. It is not only a sort of athletic training of the intellect but also a very effective learning method, because of the numerous advantages problem solving offers over ordinary descriptive teaching. It allows a much more active participation of students in any instructional program,[1] something that has been shown recently to improve the efficiency of information transfer.[2] In addition, problem solving may more effectively arouse the student's interest in a particular subject, which is, needless to say, an indisputable advantage for the learning process. Furthermore, the complex mental mechanism[3] associated with the intensive search for solutions to given problems enhances those abilities related to critical thinking.[4] This does not invalidate recall thinking or what education scientists call cognitive memory. However, the classical teaching process seems to have put too much emphasis in the latter, to the point of completely ignoring formal courses of problem solving techniques. This is perhaps the consequence of the upsetting effect of the vast amount of information every scientist has to cope with every week, and feels obliged to pass on to students. The emphasis, it seems, is on information courses, leaving aside the formative aspects of teaching.

However, in essentially all branches of chemistry, problem solving is also a basic requirement for professional practice. Technicians in industry, on the one hand, frequently confront situations whose solutions cannot be found in manuals and training booklets. Sometimes a delay in the application of the right strategy may have disastrous results: loss of feedstock, time, energy, or even the chemical plant itself. Executives also are not exempt from the risk of being unable to make a decision on the basis of the analysis of a given scenario.

Scientists, on the other hand, are sophisticated problem solvers. Not only are the proposing of hypotheses, the designing of experiments to prove or disprove them, and the drawing of valid conclusions from correlation of data complex intellectual processes that require much more than imagination and systematics, but the capacity to interpret the often circuitous and intricate data nature offers as a response to our keenly designed experiments is a challenging daily situation in the successful scientific laboratory.

It is not surprising, therefore, that interest in the study of problem solving is showing signs of renewal in recent years, and is slowly entering the classroom.[5] A much more serious effort will have to be made, however, to achieve a satisfactory level. In fact, at present all we regularly see in graduate level chemistry is students continuously challenging their comrades with the most fiendishly complex mechanistic problems they are able to pick up from the latest issues of chemistry journals. However, there are no regular courses on problem solving. One may be misled by the sight of students surrounded by coffee mugs late at night scrambling to find ways to explain a convoluted reaction mechanism on scraps of paper or blackboards, to conclude that these people are becoming efficient problem solvers. This is not true. A closer look will reveal that frequently their search for a solution is a disordered and confusing attempt to move electrons about until the atoms of the starting materials fall in the right place in the final product, loosely following existing chemical theories. They will never learn how to do this systematically by simply challenging each other.

The teaching of how to focus problem solving properly, so strongly emphasized in efficiency-demanding industrial environments and so little stressed in chemistry courses, might bring some balance in the formation versus information dilemma. In addition, problem solving techniques are intimately related to *analysis* and *synthesis*, two mainstays of human evaluative thinking.

Organic chemistry is there to help. The number of structures possible for a small set of carbon atoms and their means of conversion to other materials is awesome. Yet they all follow certain rules that cannot be ignored. The combination for focusing problem solving therefore seems perfect: A very large set of possibilities, situations and scenarios such as structures, reagents, and reaction conditions, all related to a number of rules—the laws and principles of chemistry. Consequently, properly focused organic reaction mechanisms provide a very powerful tool for learning the intricate process of situation analysis, correlation analysis of available data, and the development and selection of proper criteria for choosing the correct answer.

This means that in order to communicate the *know how* and *know why* of problem solving effectively using organic chemistry mechanisms it is imperative that the rules of the game be known first; that is, the student should have a good grasp of organic reactions in general, of stereochemistry, and of physical organic chemistry. Although the latter is a ''remarkably ill-defined subject'' in the words of Professor N. S. Isaacs,[6] owing to the vast number of topics covered under that heading, its principles must be kept in mind whenever one is confronted with a reaction mechanism in order to avoid falling into the attractive field of open speculation. Of course, it is vital to be well acquainted with any particular area of knowledge in order to be able to detect the limits of the application of its principles. In terms of organic chemistry, the reader had better be familiar with reaction rates, isotopic and solvent effects, deuterium and other tracer labeling, and the trapping of reactive intermediates, in order to know when to use these powerful techniques and their underlying principles when one is left alone with one's wits trying to figure out a mechanism.

Besides being highly entertaining, the search for solutions to mechanisms embodies a highly profitable educational process of orderly thought involved in handling several concepts of varying complexity centered around a scheme. Similar to the ladder that allows one to look over the trees to see the forest, problem solving helps to develop the ability to find relationships between seemingly disparate topics. It enables us to differentiate the important from the futile, the grain from the chaff. It helps us to identify, at an early stage of development, important research subjects that may eventually evolve into relevant fields of study. This lightens appreciably the heavy burden of keeping pace with an ever expanding chemical literature. Analytical reasoning embodied in the techniques of problem solving is also a powerful aid in the design of experimental methods in new areas of research. Many reports are in effect the result of imaginative postulates put forth on a rigorous analytical basis. Finally, mechanistic problems are an excellent means of introducing named reactions or processes whose theoretical basis can be best understood by means of a good example. This particular feature has been illustrated repeatedly in many of the problems presented in this book.

It must never be forgotten, however, that the key to success is more in the hands of the reader than in the quality of the most brilliant of texts.

REFERENCES

1. S. Markle, *Designs for Instructional Designers*, Stipes, Champaign, IL, 1978.
2. R. P. Steiner, *J. Chem. Ed.*, **57,** 433 (1980).
3. H. G. Elliott, *J. Chem. Ed.*, **59,** 719 (1982).
4. M. J. Pavelich, *J. Chem. Ed.*, **59,** 721 (1982).
5. See, for instance, A. D. Ashmore, M. J. Frazer, and R. J. Casey, *J. Chem. Ed.*, **56,** 377 (1979); D. S. Bree, *Instruct. Sci.*, **3,** 327 (1975); J. G. Greeno, ''Process of understanding in problem solving,'' in *Cognitive Theory*, N. J. Castellan, D. B. Pisoni, and G. R. Potts, Eds., Vol. 2, Wiley, London, 1977.; R. E. Mayer, *Thinking and Problem Solving, An Introduction to Human Cognition and Learning*, Scott, Foresman, Glenview, IL, 1977; A. Newell and H. A. Simon, *Human Problem Solving*, Prentice-Hall, Englewood Cliffs, NJ, 1972; W. A. Wichelgren, *How to Solve Problems*, Freeman, New York, 1974.
6. N. S. Isaacs, *Reactive Intermediates in Organic Chemistry*, Wiley, New York, 1979.

THE RULES OF THE GAME

Any self-developed method or discipline will consist of personal features and individual points of view that may or may not reflect the opinion of the majority which may not necessarily be the most valid position. One should expect, therefore, that the criteria of those readers who have already created their own systematic approach to problem solving in mechanistic organic chemistry may be in conflict with some sensitive points expressed here. In addition, there are those who feel that rules in general jeopardize their ability to create and thus to find original solutions to problems.

What follows, nevertheless, is neither a set of restricting and general laws, nor anything vaguely resembling the Philosopher's Stone. Rather, it is a collection of suggestions and techniques that have proved to be quite successful for me in problem solving as applied to organic chemistry. Those readers who will dissent from these ideas are invited to write their own book on this virtually untapped field.

I. SLOW MOTION, DEEP INSIGHT

Modern civilization forces us to live, move, sleep, eat, and think as fast as we can, as if the human being were nothing but a sort of unmatched computer. Not so, for the computer is, as of this minute, a dumb black box incapable of intuition. These machines serve us admirably well and at astonishing speeds, supplying us with reams of data, numbers, and calculations, but they are nothing but tools. However, we still do the thinking, press the keys, and design the programs. So, leave the speedy portion of your work to those technological wonders and dedicate yourself to careful thinking. Problem solving demands a great deal of mind churning, and unless the problem is well ruminated, mental indigestion may result. Although attentive deliberation is a time consuming operation, at the same time, it is an exceedingly rewarding one. Enjoy the chemistry of your problem much in the way an artist savors the embroidered details of his craft. One stands to learn a great deal more by solving a single problem in depth than by cranking out 10 shallow answers to as many problems in the same period of time.

In addition, it is to one's advantage to forget altogether the existence of time if full mental concentration is desired. In fact, unless undisturbed concentration is achieved, futile daydreaming, petty distractions, and the noisy manifestations of modern culture will start working against you. For those who actually cannot ignore the march of time, they will soon realize that periods of deep concentration are by far more time efficient in terms of solution output than any other clock-regulated period of the day.

So, make yourself comfortable, surrounded by as much reference material as possible, and get started. Ah!, do not forget to unplug the telephone. Nothing is more annoying than a phone call in the middle of an exothermic reaction out of control!

II. CLEAR DRAWINGS, THREE-DIMENSIONAL STRUCTURES

If it were so ridiculously obvious and a waste of editorial space to assert that chemical structures should always be drawn as clearly as possible to simplify figuring out a problem, there would not be so many midnight horror stories of teachers and scholars trying to untangle the riddle of cumbersome molecular drawings buried under hesitating arrows and erasures while correcting organic chemistry tests.

Well shaped and elegant pictorial representations of one's ideas usually have the beneficial feedback effect of driving away the fog that surrounds vaguely stated contentions, clearing up thoughts, and even providing leads to possible solutions.

In addition to looking at clear diagrams, it is also advisable to look at compounds from several angles, not just from the most obvious. This helps in the identification of reactivity patterns other than the usual ones and in recognizing a familiar fragment in a complex structure, a feature that might well be the key to a whole mechanistic scheme. For example, it will take a little longer to visualize the indole portion of a complex alkaloid such as **I** for those who follow the classical portrayal of indole with the nitrogen atom facing down on the right-hand side of the molecule, than for those who draw this compound upside down, tilted, or sideways.

The three-dimensional drawing of organic molecules, although somewhat difficult, is particularly helpful in visualizing relationships between interacting groups, steric encumbrance, conformational changes, and even the very feasibility of a reaction pathway. The two-dimensional or flat representation of the intramolecular displacement reaction that leads to a compound with the very appropriate name Twistanone (**III**)[1] may be thought quite absurd owing to the seemingly long distance that lies between the two reacting sites (see Scheme 1). The three-dimensional drawings **IV** and **V** of its precursor **II** show, however, that the two carbons involved interact at close range. In addition, the uncomely slash across structure **III** is nothing but a carbon–carbon bond nearly indistinguishable from the others.

Looking at molecules from different angles in their spatial picture is also a very effective means of foreseeing chemical transformations as well as simplifying drawings. Some accessible computer programs perform wonders in spinning complex molecules in space. However, doing this with pencil on a piece of paper is even more entertaining and challenging. It may take, though, some training and ability to turn an imaginary complex surface such as a molecular model around in space by manipulating it on the flat surface of a sheet of paper.

For many of the readers who glanced rapidly at Scheme 1, it may not become apparent until a second thorough examination that structures **VI–IX** all represent the very same Twistanone (**III**). Each one of these drawings seems to have a character of its own, depending on its visual relationship with more familiar molecular systems. Thus while **VI** is just a twisted maze of strained

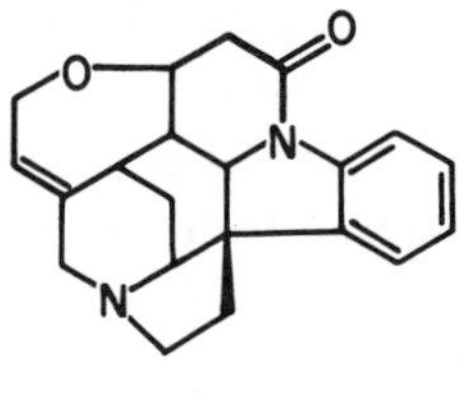

I

(STRYCHNINE)

R = $CH_3SO_3^-$

SCHEME 1

bonds, these stresses magically disappear in the enchanting symmetry and topological beauty of structure **VII.** This harmony may be subconsciously related with chemical stability. However, structures **VIII** and **IX** permit the association of Twistanone with the versatile bicyclo[2.2.2]octane ring system, from which a more complete reactivity diagram may be conceived.

The design of synthetic trees can also be profoundly influenced by the particular orientation we choose for the target compound. Congressane, for example, a pentacyclic compound that owes its odd name to having its structure used by an imaginative artist as an emblem of the 19th International Congress of Pure and Applied Chemistry (1963) (where attending chemists decided then to undertake its challenging synthesis), can be drawn in many ways. Four of these are portrayed here. While **X** and **XI** are artful representations more linked to shield decorations of African warriors in Dr. Livingstone's time than to any reactivity pattern or synthetic origin of Congressane, structure **XII** dutifully shows two cyclohexane chairs linked by two kinds of bridges with clearly defined equatorial and axial positions. Were we to depict Congressane as **XIII,** namely, an adamantalog of adamantane (**XIV**),[2] our vision of its synthesis would probably be oriented quite differently because the target compound now be-

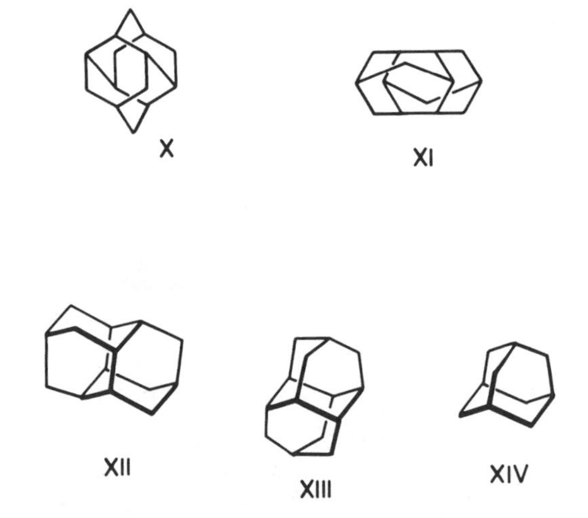

comes one cyclohexane chair with six axial methylene substituents that converge towards two methines situated at the apexes of the molecule.[3]

If the reader still had trouble in visualizing **XV** and **XVI** as different pictures of the same compound after the first minute or so, some more practice may be needed before this helpful technique can be of any use.

III. PRESERVE ATOMS IN PLACE; KEEP THINGS SIMPLE

Unfortunately, clear drawings are not all that is necessary in dealing with complex reactive intermediates. It is also essential to carry out bond dislocations, reconnections, and rearrangements. In doing this it is advisable to keep all atoms in place, particularly those belonging to the main molecular skeleton, while performing bond changes in single steps, even if the result is an apparently strained system. Once the structure of the new intermediate is well understood, then it may be graphically changed into something more pictorially appealing

XV

XVI

as described previously, and the mechanistic route may proceed from this new structure.

For example, the conversion of triene **XVII** into the vinyl acetate derivative[4] **XVIII** requires some bond reorganizations that may be difficult to visualize (see Scheme 2). Individual steps will be easier to contrive and understand by keeping those atoms and bonds not involved in the particular single operation in place. The combination of fixing atoms and bonds and later changes to realistic pictures gives a maximum of transparency and elegance to the sequence. Compare, for instance, routes **A** and **B** of Scheme 2. Atoms remain fixed in pathway **A** while **B** is a more realistic representation of the intermediates involved.

In order to simplify reaction sequences further it is convenient to ignore all of that sector of the molecule that does not undergo any change or that does not participate in the particular step one is dealing with. This is especially helpful in handling large molecules such as steroids, porphirins, and complex alkaloids. Otherwise these elaborate drawings will be distracting, time consuming, and usually have an overwhelming effect on one's hopes of finding a solution. In reviewing Woodward and Eschenmoser's famous synthesis of cobyric acid and Vitamin B-12, one cannot resist a sense of awe caused by the gigantic, convoluted intermediates. However, most of the synthetic operations in the sequence, in all their beauty and experimental difficulty, are conceptually speaking within the reach of the good graduate student of organic chemistry.

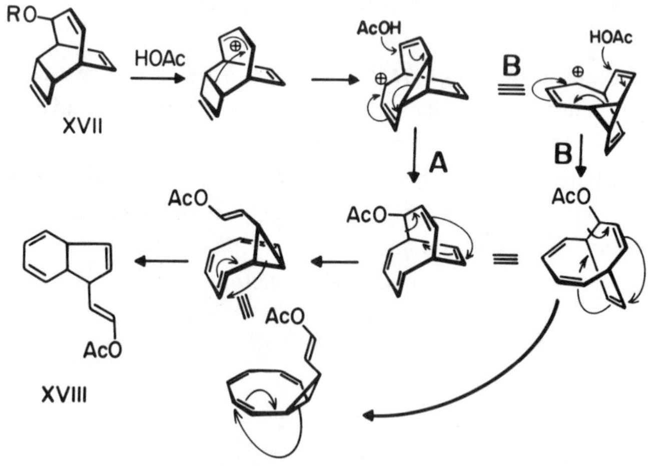

SCHEME 2

IV. ATOM AND BOND BUDGETS

The standard procedure of comparing empirical formulas of starting materials and products usually provides hints about whether the reaction under scrutiny is an isomerization, a fragmentation, a substitution, or an addition process. It also yields information on the possible nature of the ejected fragment if there is one, since it is normally much smaller than the parent molecule, hence easier to reconstruct from an empirical formula.

For instance, that an isomerization followed by decarboxylation is the most probable process involved in the transformation of keto ester **XIX** into ketone[5] **XX** is suggested, on the one hand, by the fact that the four-carbon chain of the butenyl ester is present in the product ketone on the same position occupied by the carboxylate unit in **XIX.** On the other hand, the empirical formulas indicate the departure of one carbon and two oxygen atoms, in all probability belonging to carbon dioxide.

At times useful clues can be derived from carbon-to-carbon and carbon-to-heteroatom bond counting in starting materials and products. The difference might shed some light on the number of new bonds that one has to construct or destroy to account for the product. This, in turn, could suggest possible reaction pathways and relate the problem to some known reaction.

The interesting cyclization[6] that yields two fused rings in one operation from the open chain diene **XXI** entails the consolidation of two new C–C bonds with the sacrifice of no visible fragment—not an intramolecular substitution, therefore—but of two double bonds. This is a clear suggestion of an intramolecular Diels–Alder cycloaddition.

V. OXIDATION LEVELS AND MECHANISM

Relative oxidation levels of products and starting materials as well as that of putative intermediates is an extremely valuable tool in the assessment of reac-

NaH, Me_3SiCl

XIX → XX

XXI → (Δ) XXII

tion mechanisms. However, the level of oxidation of any given organic material is much more difficult to evaluate than that of an inorganic compound, where a cautious count of valence electrons will suffice in most cases. It is also unfortunate that the well established concept of oxidation number may bring havoc and confusion rather than a remedy to our evaluation of oxidation levels, since the oxidation state of each individual carbon depends on the oxidation number of neighboring carbons and often this leads to fractional numbers.

There are two ways to solve this problem. The first is the use of a classical unsaturation number that can be obtained from the empirical formula, as one is taught in freshman chemistry courses. Comparison of the atomic composition of starting materials and products yields the unsaturation number balance that one needs to assess oxidation level changes as a whole. Frequently, however, it is also useful to learn the way by which parts of the molecule undergo oxidation level changes in order to concentrate one's attention on that particular portion of the compound. This leads us to discuss the second way of solving the oxidation level problem.

Think, for a moment, about the principal role played by the central carbon of a functional group in forming C–C bonds. Then ask how many C–C bonds that particular carbon is capable of forming until it gets rid of all heteroatoms directly substituted on it as well as unsaturations. This number could be anything from a maximum of four, for stable species of carbon, to a minimum of zero in the case of nonactivated hydrocarbon methylenes and methane. Then give this number a name, say Functionality Number, abbreviated as FN.

Now, let us take a simple functional group such as an OH and substitute it on our carbon atom. The carbinol we now have is capable of yielding only one C–C bond by way of substitution or one unsaturation on our carbon by means of elimination. Alcohols should then be classified under the FN = 1 group. Other oxygen based substituents such as ethers, carboxylic and sulfonic esters, and other heteroatom-containing derivatives such as amines and sulfides behave similarly, as long as we have only one of the heteroatoms substituted on our central carbon. By the same token, alkyl halides, diazonium salts, and even the carbenium ion itself are all capable of giving only one C–C bond with carbon nucleophiles, but never two. It seems, therefore, that all these functions fall in the same FN category, that of FN = 1.

If we put two of these functions on the same carbon, in principle, one may be able to produce a maximum of two C–C bonds in reactions with appropriate carbon nucleophiles. Which functions are these? Acetals, thioacetals, aminals, α-dihalides, and so on. Then, all of them should be FN = 2 groups. This analogy continues with triheterosubstituted (FN = 3) and tetraheterosubstituted (FN = 4) carbon functions. Unsaturated systems are also classifiable in these terms. Every one knows that a ketone can yield two C–C bonds. This circumstance evolves into the realization that the π and the σ bonds of C=X both serve their individual purpose as potential precursors of C–C bonding. This allows the inclusion of imines, thioketones, and diazocompounds as functions capable of forming a maximum of two (FN = 2) C–C bonds. Similarly, esters, imino- and thioesters, and acyl halides display central carbons with FN = 3, while carbonates, carbamates, ureas, and carbon dioxide are all FN = 4 groupings.

All the groups mentioned have the central carbon acting as an electrophile for C–C bond forming reactions, so a positive sign is given to the corresponding FN number. This leaves room for including functions with nucleophilic forms of carbon with negatively signed FN's. Thus, Grignard reagents and equivalent forms of negatively charged carbon appear in the FN = −1 group. Carbanions are also more conveniently handled in other more stable, covert forms. These forms involve the β carbon of ketone enolates, enol ethers and all other structures with the general composition R—C=C—X, where X is an electron donor (+R) heteroatom.

There are rare cases of a carbon acting twice as a nucleophile or electron donor in the construction of C–C bonds: then FN = −2. Under this heading, only a handful of rather active and unstable functions such as the recently reported β-metallated enol ethers and enamines can be grouped. Their use is restricted to systems in which the much more acidic proton on the α carbon is blocked by substitution. A more common function for the FN = −2 group is the phosphorous ylide, better known as the Wittig reagent, for making terminal alkenes from ketones and aldehydes. Finally, the recently introduced α-sulfone dianion[7] would be a genuine representative of the FN = −2 group.

Placing all the functions mentioned previously in a table, ordered according to their FN classification, gives an orderly collection of groups as in Table 1. Other groups have been included without individual discussions for the sake of brevity. It is quite simple, however, to figure out why they are where they are.[8]

A number of useful relationships between functionalities may be derived

TABLE 1. Classification of Common Functional Groups According to Their Functionality Numbers.

FN	-2	-1	0	1	2	3	4
	$R{-}\overset{=}{C}{-}SO_2R'$ $2Li^+$	R_3C^-	$R_3C{-}H$	$R_3C{-}OR'$ $R_3C{-}NR'$ $R_3C{-}SR'$ $R_3C{-}X$ $R_3C{-}\underset{+}{N}{\equiv}N$	$R_2C{-}(OR')_2$ $R_2C{-}(NR'_2)_2$ $R_2C{-}(SR')_2$ $R_2C{-}X_2$	$RC{-}(OR')_3$ $RC{-}(NR'_2)_3$ $RC{-}(SR')_3$ $RC{-}X_3$	$C(OR)_4$ $C(NR_2)_4$ $C(SR)_4$ CX_4
				$R_2C{-}O(NR',S){-}CR_2$ (ring)	$R_2C{-}O(NR',S){-}C(R)(OR')$ (ring)	$R'{-}\overset{O}{\overset{\Vert}{C}}{-}OR\ (NR_2, SR)$ $R'{-}\overset{NR''}{\overset{\Vert}{C}}{-}OR\ (NR_2, SR)$	$(R_2N,RS)RO\overset{O}{\overset{\Vert}{C}}OR'(NR'_2,SR')$ $(R_2N,RS)RO\overset{NR''}{\overset{\Vert}{C}}OR'(NR'_2,SR')$
	$R_2C{=}PR'_3$		$R_3C{-}CR_3$	R_3C+	$R_2C{=}O$ $R_2C{=}NR'$ $R_2C{=}S$	$R'{-}\overset{S}{\overset{\Vert}{C}}{-}OR\ (NR_2, SR)$ $R{-}\overset{O}{\overset{\Vert}{C}}{-}X$	$(R_2N,RS)RO\overset{S}{\overset{\Vert}{C}}OR'(NR'_2,SR')$ $(R_2N,RS,RO,X)\overset{O}{\overset{\Vert}{C}}X$
					$R_2C{=}\underset{+}{N}{=}\underset{-}{N}$	$R_2C{=}C{=}\underset{+}{N}{=}\underset{-}{N}$	
	$R{-}\bar{C}{=}CR'{-}OR''$ $R{-}\bar{C}{=}CR'{-}NR''_2$ $R{-}\bar{C}{=}CR'{-}SR''$	$R_2C{=}CR'{-}O^-$ $R_2C{=}CR'{-}OR''$ $R_2C{=}CR'{-}NR''_2$ $R_2C{=}CR'{-}SR''$		$R_2C{=}CR_2$	$R_2C{=}CR'{-}OR''$ $R_2C{=}CR'{-}NR''_2$ $R_2C{=}CR'\ SR''$ $R_2C{=}CR'{-}X$	$R_2C{=}C(OR')_2$ $R_2C{=}C(NR'_2)_2$ $R_2C{=}C(SR')_2$ $R_2C{=}CX_2$	$O{=}C{=}O$ $RN{=}C{=}NR'$ $S{=}C{=}S$
		$RC{\equiv}C{-}OR'(NR'_2, SR')$			$R_2C{=}C{=}CR'_2$ $RC{\equiv}CR'$ $R_2C{=}\overset{+}{C}R'$	$R_2C{=}C{=}O(NR'_2,S)$ $RC{\equiv}C{-}OR'(NR'_2,SR')$ $RC{\equiv}C^+$	
						$N{\equiv}CR$	$N{\equiv}C{-}OR(NR_2,SR)$ $N{\equiv}C{-}X$

from this table. In its construction we have performed only two basic operations: (1) to replace C–R and C–H bonds progressively by more active C–X linkages (X = O, S, N, halogens) using only those heteroatoms most commonly found in organic chemistry, and (2) to introduce π bonding gradually. This progressive construction of one function from another by the addition of always the same type of component in each step provides a good basis for drawing analogies in the chemical behavior of the functions included in the sequence. For example, replacing R by OR in a ketone yields an ester. Substitution of a second R by OR affords a carbonate. Yet all these functions maintain the central carbonylic section essentially unchanged, and consequently share a number of chemical properties. Furthermore, if the introduction of other heteroatoms, such as in the transformation of a ketone into an imine or thioketone, do not involve a deeply rooted alteration of the chemical behavior of the central carbon—differences are more of degree than of form—then the ketone family may be expanded considerably to include the iminoester, imidazole, and thiocarbamate functions. Along the same lines, allenes, ketenes, carbodiimides, ketenimines, and carbon dioxide are closely related functions.

These relationships make the handling of uncommon functionalities considerably easier, in particular those of high heteroatom content. In addition, the design of new, still unexplored functions by extrapolation of data in Table 1 is also conceivable. The FN = 2 equivalent of the carbenium ion, for example, the well-known vinyl cation, was a theoretical entity some decades ago, much as its FN = 3 equivalent, the alkynyl cation, is today. In fact, this mechanism of analogy is an effective means of designing research to pierce the unknown and the orderly collection of functional groups may bring the necessary leads to construct such an analogy .

The relative position of a given group in Table 1 is also linked to its relative oxidation level, which was the starting point for the development of this table. It will be easy to realize that members of the same FN group do have the same oxidation level. Ketones, aldehydes, imines, acetals, diazo compounds, and enolates have the central carbons with the same oxidation level. By the same token, ortho carbonates, carbon dioxide, carbon disulfide, phosgene, and carbonates also share a common oxidation level. This means that the functions contained in the same FN class may be interconverted without the use of reducing or oxidizing reagents.

What about functions belonging to different FN groups? Take, for example, an alcohol and a ketone. Everyone knows that an oxidation is required to transform the former into the latter. Moreover, this is a one-step oxidation, namely, one in which only two electrons are transferred. The difference of FN value between alcohol and ketone is one unit. Consequently, each FN unit will correspond to a one-step difference in oxidation level. Alcohol and ester functions are separated by two FN units, which duly reflects their four-electron distance (two oxidative steps). All this means that: *The conversion of a carbon atom to a higher FN will represent a formal oxidation, and an FN change in the opposite direction will be a reduction.*

This is important information to have at hand when handling functional groups. It is particularly useful if we recall, in addition, that the process of C–C bond formation is essentially a reductive operation, whereas the separation of a C–C bond entails an oxidation of the remaining positively charged carbon. For example, methyllithium will convert an ester (FN = 3) to a ketone (FN = 2) under controlled conditions, or to a tertiary alcohol (FN = 1) under more severe conditions by successive C–C bond forming processes. Conversely, acetone will be transformed by pyrolysis into a ketene (central carbon FN = 3) by α fragmentation.

We may conclude, then, that functionality numbers of separate sections of the molecules of starting material and products give a much more informative picture of sectional oxidation level changes in reaction sequences than comparison of empirical formulas, and is thus a valuable tool in the assessment of reaction mechanisms. We shall come back to this in the subject that follows.

VI. MOLECULAR DISSECTION: WORKING THINGS BACKWARDS

Most people struggle to work out a mechanism by pushing forward from the starting material the best way they can, because there is a general feeling that it is the starting material and not the product that holds the key to the solution. This is dead wrong. For one thing, most compounds embody a number of visible features in their structures that may suggest possible entries into their synthesis, thus allowing for the design of a synthetic sequence in a backward fashion.[9]

In addition, reaction products display in most cases the clues one needs to explain their mechanism of formation. The problem then becomes how to identify those magic features. For this purpose fragmentation analysis is positively the method of choice.

The fragmentation technique consists of the dissection of the target molecule into its composite parts. The basic principle of this bond clipping process is to search among the separated fragments for those molecular bits that resemble the composite parts of the starting materials: a carbon chain here, a functional group or substitution pattern there, and so on. From the structure of these fragments it becomes considerably easier to figure out the behavior of the starting material as a reactive species and consequently the nature of the chemical transformations required to arrive at products.

Dissection is best approached in a progressive way, selecting first the most obvious sections. Then the remaining parts will reveal themselves either as fragments of the starting materials or the result of rearrangements that take place in the course of the mechanistic sequence.

The jawbreaking name of compound **XXV,** let alone its IUPAC denomination, will surely be more difficult to pronounce than finding a reasonable mechanism for its construction from **XXIII** and **XXIV,**[10] particularly if one applies its molecular dissection in terms of the starting materials. First, segments **A** and **C** obviously correspond to 2 mol of diester **XXIII.** The relationship of this compound to segment **B** is a little less obvious, but carries with it the story of an aldol condensation revealed by the tertiary alcohol unit. Finally, fragments **D** and **E** show rather obscurely that, being two-carbon units, they probably are derived from the bis-formaldehyde **XXIV.**

XXIII XXIV XXV

E = $COOCH_3$

A much deeper insight into reaction mechanisms can be achieved by fragmentation analysis; that is, not only do the composite parts become visible, but it also enables us to see the actual mechanistic logic involved. This calls for interpretation of the sequence of events in a backwards fashion. For those readers who judged this problem too simple, here is a more demanding transformation.[11]

Three cyclization steps constitute a remarkable feat for any compound in just a one-pot operation. Three, in fact, because there is nothing in the product that indicates the presence of the cyclohexadione system of the starting material, at least in its original form. The massive change undergone by **XXVI** can be rationalized best by fragmentation analysis. One version of the various possible fragmentations follows.

First, the lactone fragment **A** (see Scheme 3) may result from the *O*-acylation of a ketone in its enol form **A′**. Second, the enone moiety represented by segment **B** conceivably may be derived from an intramolecular aldol condensation followed by elimination of water, just as in the Robinson annullation process. In turn, the 3-pentanone chain that is needed for this step is clearly discernible in the keto side chain that stems from the quaternary center of **XXVI.** Therefore, it should be easy now to detect fragment **B** in the starting material. The identification of the origin of parts **A** and **B** in turn allows for the construction of part **C** as a cyclopenta-1,3-dione (**C′**) in which the *upper* ketone must have come from the ester unit in order to maintain the position of the quaternary carbon (denoted C*), whereas the ester side chain is congruent with the remaining carbons of the original six-membered ring (see C″).

Furthermore, the fragmentation of the six-membered ring of **XXVI** that obviously must be taking place can be construed on several accounts (not indicated). In order to achieve this, it is crucial to take into account the transformation of the quaternary carbon C* into a tertiary center. It is also significant that a new C–C bond is being constructed from C**.

COOEt

PPA

XXVI XXVII

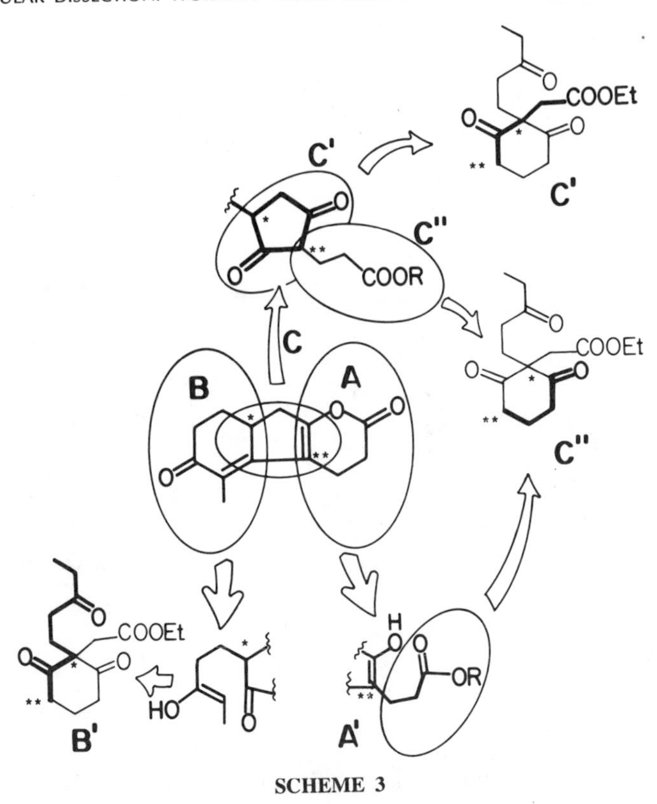

SCHEME 3

Is it still necessary to portray the proposed mechanism in full? Yes, perhaps for the sake of didactics we should (see Scheme 4). Little additional comment will be required though. The fragmentation of the 1,3-cyclohexadione ring mentioned would occur under the auspices of polyphosphoric acid as shown in **XXVIII.** The new carbon–carbon bond shared by C** and the ethyl ester group would be constructed between the kinetic enol form of **XXX** and the carbonyl carbon five atoms away. The other two cyclization steps that follow from this have already been uncovered by our previous fragmentation analysis.

Fragmentation analysis can also be used in combination with the Functionality Numbers defined previously (see Section V). In some cases, the identification of the starting material in the composite parts of a given product is difficult, requiring an above average imagination. Here we apply FN. If the molecular splitting entails C–C bond rupture, one might perfectly well find out which were the functional groups required to reconstruct these bonds by as-

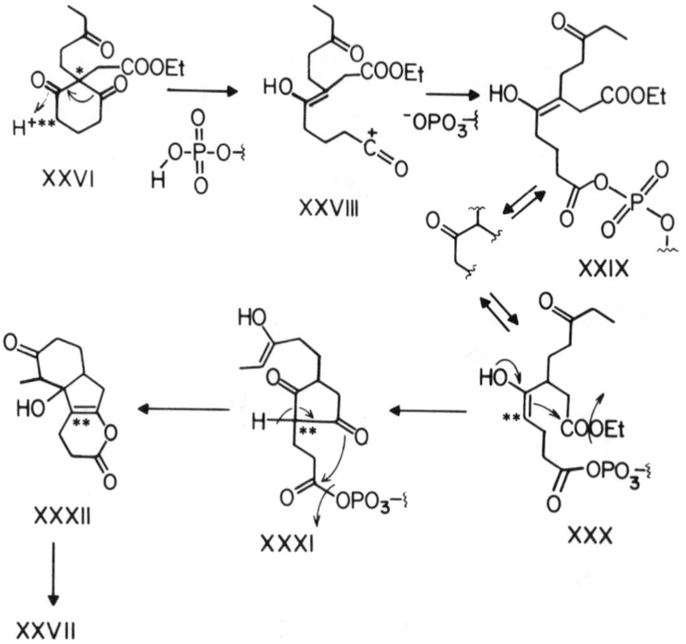

XXVI XXVIII XXIX XXX XXXI XXXII XXVII

SCHEME 4

signing FN's to the fragments. These functional groups would be chosen from Table 1. The molecular pieces with added functions would then resemble the starting materials more closely than the bare fragments themselves.

One interesting example is provided by transformation of the vinyl chloride derivative **XXXIII** into **XXXIV**.[12] No matter how one takes the product apart, the result will continue to be puzzling.

However, if this fragmentation exercise is performed with the addition of some FN information, the picture becomes much clearer. First, the unraveling of the central bonds of the cyclobutane ring yields two cyclopentane units with two carbons belonging to the FN = 1 group, but with opposite signs. A C=C

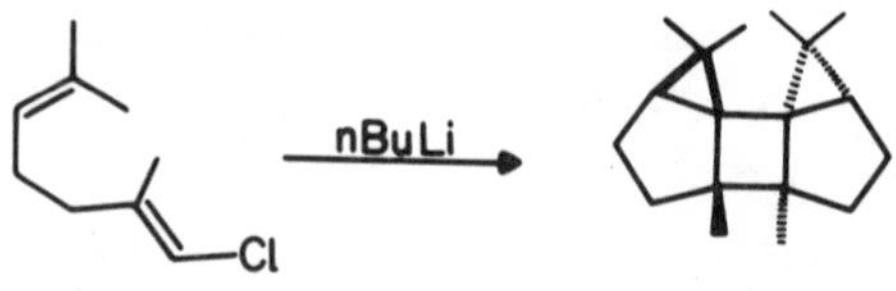

nBuLi

XXXIII XXXIV

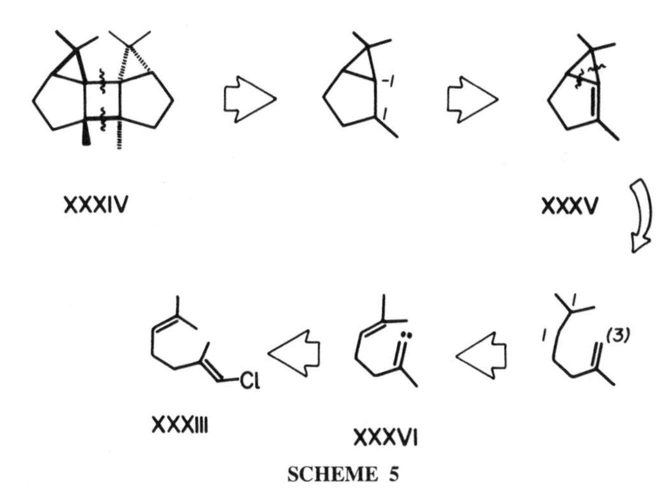

SCHEME 5

bond shared by these two carbons would be ideal to fulfill the requirements represented by these FN's and would serve as well for the [2+2] cycloaddition that would reverse **XXXV** to the product **XXXIV** (see Scheme 5). In turn, disconnecting the two =C—C bonds of cyclopropane in **XXXV** reveals the presence of an isobutenyl group and a FN = 3 carbon at the other end of the chain (one FN unit for each cyclopropyl C–C bond plus another FN unit on account of the double bond). However, the starting material features a FN = 2 carbon at this position. Our list of FN = 3 groups that include a double bond is rather short (see Table 1) and among the few candidates the vinyl carbene (not shown in the table) appears as the best suited group.

Consequently, a mechanism that accounts for this curious transformation most probably starts with the α elimination of HCl to give the alkylidene carbene, which then adds across the trisubstituted double bond at the opposite end of the molecule. Finally, the intermediate alkene undergoes thermal [2+2] cycloaddition to give **XXXIV.**

Molecular dissection has proved to be a most powerful and convenient technique to unveil complex mechanistic sequences. We shall appeal to its usefulness repeatedly in the following problems.

REFERENCES

1. See, among others, H. W. Whitlock, Jr., *J. Am. Chem. Soc.*, **84,** 3412 (1962); A. Belanger, Y. Lambert, and P. Deslongchamps, *Can. J. Chem.*, **47,** 795 (1969).

2. V. Prelog, *Pure Appl. Chem.*, **6,** 545 (1963); G. M. Blackburn, D. W. Cameron, A. R. Katritzky, and R. H. Prince, *Chem. Ind.* (*London*), 1349 (1963).
3. In the form presented by **XIII,** Congressane would be the smallest component of the diamond lattice, since diamond is an infinite adamantalog of cyclohexane, See C. Cupas, P. von R. Schleyer, and D. Strecker, *J. Am. Chem. Soc.*, **87,** 917 (1965).
4. J. T. Groves and C. A. Bernhardt, *J. Org. Chem.*, **40,** 2806 (1975).
5. R. M. Coates, L. O. Sandefur, and R. D. Smillie, *J. Am. Chem. Soc.*, **97,** 1610 (1975).
6. R. F. Borch, A. J. Evans, and J. J. Wade, *J. Am. Chem. Soc.*, **97,** 6284 (1975).
7. A. Bongini, D. Savoia, and A. Umani-Ronchi, *J. Organomet. Chem.*, **112,** 1 (1976). [(Phenylsulfonyl)methylene]-dilithium reacts with α dihalides, haloepoxides, halocarbonyls, dicarbonyls, halonitriles, and α,β-unsaturated carbonyls to give carbocycles in good yields. See J. J. Eisch, S. K. Dua, and M. Behrooz, *J. Org. Chem.*, **50,** 3674 (1985).
8. For further details see M. E. Alonso, *J. Chem. Ed.*, **54,** 568 (1977); *Acta Cient. Venez.* **35,** 317 (1984).
9. For an excellent text about the disconnecting approach to organic synthesis see S. Warren, *Designing Organic Synthesis; a Programmed Introduction to the Synthon Approach*, Wiley, New York, 1978.
10. K. C. Rice, *Tetrahedron Lett.*, 3767 (1975).
11. A. M. Chalmens and A. J. Baker, *Tetrahedron Lett.*, 4529 (1974).
12. G. Kobrich, *Angew. Chem. Int. Ed. Engl.*, **12,** 464 (1973).

PROBLEMS AND DISCUSSIONS

In the first chapter a number of techniques useful for the elucidation of probable reaction mechanisms were briefly described. However, practice, rather than mere description, may be the best means of gaining dexterity in the performance of any difficult art. The collection of problems and their discussion that follows is intended to give ample opportunity to exercise these and other theoretical considerations by providing a number of examples designed to fix in the reader's mind the basic principles of mechanistic reasoning for organic reactions.

Fifty-six problems, arranged from the simple to the more complex, chosen from current chemistry journals, constitute this collection of exercises. The criterion of arranging the problems according to their complexity was chosen to avoid overwhelming the students with unnecessarily difficult, inaccessible problems at an early stage and to give them the opportunity to gradually develop their own skills in the mastering of problem solving.

The index at the end of the book has been designed to provide a means of quickly finding those particular reactions and processes of interest and as an effort to make up for those deficiencies thay may be present in the problem ordering scheme used here.

Transformations featuring concepts of general relevance were selected for the most part instead of strange and rare chemical phenomena of limited interest to the general audience, so as to illustrate widely accepted chemistry of current interest, for the purpose of stimulating instructive discussions. More than one answer is often portrayed, and frequently several alternatives to the original proposition, by the authors of the reaction under scrutiny, are given. Not all of these proposals will be valid, of course, and some may even seem a little strained. However, it must be kept in mind that the behavior of matter at the molecular level is extremely complex and current chemistry enables one to understand only a fraction of this complexity. Unsatisfactory proposals will be critically analyzed to indicate why they are not acceptable. All temperatures, unless otherwise noted, will be given in degree celsius (°C).

In order to best use this collection of problems, it is strongly recommended that readers first work out their own solution to a given problem and then carefully analyze the solutions proposed here, one by one, with extensive perusal of the references indicated. Finally, they should make analytical comparisons of all the proposed mechanistic pathways to decide which of these is more chemically sound. Students will be surprised by the number of chemical concepts they will have been driven to review and handle.

Here is your book and remember: Do not rush, take your time and enjoy it. It was written to please, not to punish.

PROBLEM 1

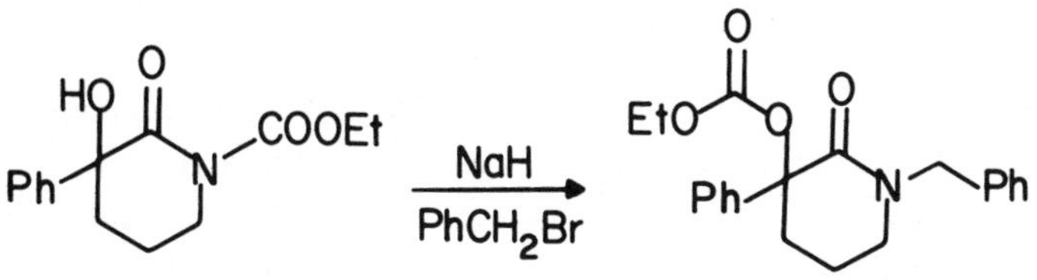

I $\xrightarrow[\text{PhCH}_2\text{Br}]{\text{NaH}}$ II

1. G. L. Grunewald and W. J. Brouillette, *J. Org. Chem.*, **43,** 1839 (1978).

PROBLEM 1
An Obvious and Yet Novel 1,4-Acyl Migration

This is nearly the simplest mechanistic problem anyone is likely to come across in an organic chemistry quiz. It involves, nevertheless, a rather unusual, albeit unhidden, 1,4 migration of the ethoxycarbonyl unit from nitrogen in the urethane **I** to oxygen in the carbonate **II.** In fact, this is the very first example ever reported of one such shift, where the acyl group is transferred seemingly intact from nitrogen to oxygen in an intramolecular fashion.

The 1,2-migration of methyl and aryl groups has been explored very extensively and convenient reviews are available.[2] In contrast, acyloxy migrations are mentioned just in passing in the chapter on neighboring group participation. Indeed, even though acyl 1,2-shifts were discovered as early as 1900[3] it took 50 years to produce just two more papers in the area,[4] with very little additional progress in recent years.[5] This is rather surprising, for it has been shown that "the migration of alkoxycarbonyl groups can be so facile as to take precedence over the possible migration of both alkyl and aryl groups."[5]

In the present case the acyl migration is easily understood if the hydroxide function becomes an alkoxide by the intervention of sodium hydride. Its approach to the carbonyl groups of the urethane unit is facilitated by its positioning in the axial conformation, as shown in **III** (see Scheme 1.1).

Of the three C–X bonds that might break in **IV,** the C–N linkage stands as the best choice, owing to the amide environment of the nitrogen atom, which experiences electron density withdrawal by the neighboring carbonyl. The subsequent incorporation of the benzylic fragment follows the course of a bimolecular nucleophilic displacement.

The mediation of a conceivably bimolecular process to account for the acyl

I → III → IV → V → II

SCHEME 1.1

group migration is unlikely in view of the lack of any *o*-benzylated product in the reaction mixture.[6,7]

REFERENCES

2. B. Capon, *Q. Rev. Chem. Soc.*, **18,** 45 (1964).
3. J. H. Ransom, *Ber.*, **33,** 199 (1900).
4. E. E. Blaise and A. Courtot, *Bull. Soc. Chim. Fr.*, **3,** 360 and 589 (1906); H. Plieninger and T. Suehiro, *Chem. Ber.*, **89,** 2789 (1956).
5. R. M. Acheson, *Acc. Chem. Res.*, **4,** 177 (1971).
6. For a related mechansim see M. D. Rozwadowska, *Can. J. Chem.*, **55,** 164 (1977).
7. For a recent example of a 1,3-acyl migration see H. L. Holland and Jahangir, *J. Org. Chem.*, **48,** 3134 (1983).

PROBLEM 2

CN, NH2, +, O, 100°, 2hr., NC, H2N, O, I, II, III (78–95%)

---

1. W. J. Nixon, Jr., J. T. Garland, and C. DeWitt Blanton, Jr., *Synthesis*, **56** (1980).

PROBLEM 2
A Classical [2+4] Diels–Alder Cycloaddition Followed by Rearrangement

It is evident that a combination of gain and loss of molecular components is involved in this transformation. Also, the fact that the vicinal methyl, nitrile, and amino groups of **I** continue to be proximal in **III** in the same order leads one to conclude that the carbon skeleton of furan **I** remains intact in the product.

It is also apparent that the oxygen atom of the amino furan derivative does not become incorporated into the product, and the stoichiometry of this reaction indicates that its loss must be taking place as water. Fragmentation analysis of the product allows the easy recognition of the starting materials, and that two C–C bonds must be formed in this reaction in a regiospecific manner (see Scheme 2.1).

There are just a few one-pot reactions in which two carbon–carbon bonds are simultaneously formed. One of these is the ancient but exceedingly useful Diels–Alder reaction, for which furans are particularly popular dienes. Consequently there would be nothing unusual about having compound **I** and methyl vinyl ketone (MVK) coupled in a Diels–Alder cycloaddition mode except for the fact that 2,5-disubstituted furans are known to be reluctant dienophiles. However, the amino group at C-2 considerably increases the reactivity of the furan nucleus as a diene by injecting electron density into the ring's π system. This phenomenon is underscored by the regiospecificity of the cycloadditon.

Furthermore, the endo epimer is expected to be formed first owing to better π-orbital overlap of **I** and MVK in the transition state. It is characteristic of oxabicyclo[2.2.1]heptanes, however, that they epimerize to the thermodynamically more stable exo isomer as shown in structures **IVa** and **IVb** (see Scheme 2.2).

O NH2 CN NH2 CN NH2 CN O

SCHEME 2.1

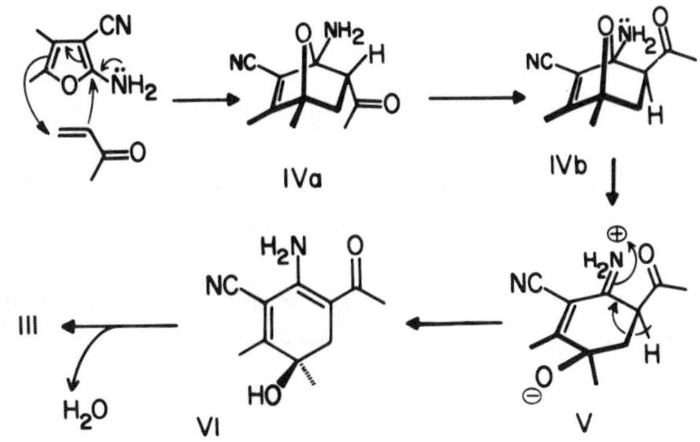

SCHEME 2.2

Adduct **IV** would be expected to have some intrinsic instability due to its α-amino ether character. In fact, this intermediate contains all the components required for its conversion to an aromatic ring. The unraveling process could well start with the intervention of nonbonding electrons of the amino group and the concomitant breakage of the oxygen bridge. Then, the ensuing alcohol (**VI**) would undergo β elimination under the driving force of aromatization to yield product **III.**

The transformation of α-pyrones to benzene and carbon dioxide upon their cycloaddition with acetylene,[2] and the reaction of 2*H*-pyrans to yield benzoic acid derivatives when exposed to acetylenecarboxylates[3] are conceptually related processes (see Scheme 2.3). These are, strictly speaking, retro Diels–Alder reactions,[4] some of which embody interesting mechanistic problems.

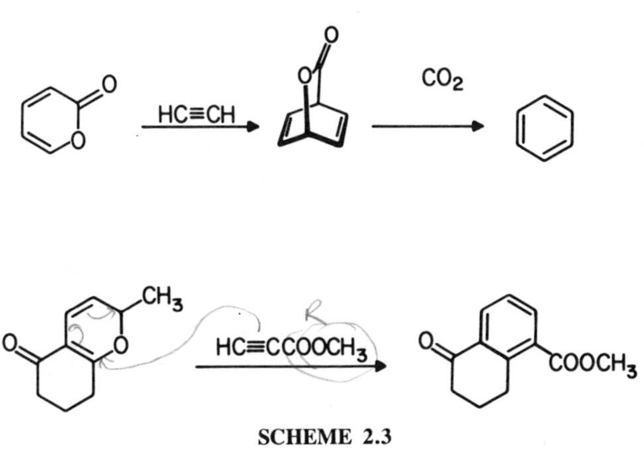

SCHEME 2.3

REFERENCES

2. G. Markl and R. Fuchs, *Tetrahedron Lett.*, 4691 (1972) and references cited therein.
3. R. G. Salomon, J. R. Burns, and W. J. Dominic, *J. Org. Chem.*, **41,** 2918 (1976); H. L. Gingrich, D. M. Rousch, and W. A. Van Saun, *J. Org. Chem.*, **48,** 4869 (1983).
4. See, for example, W. Rastetter, *J. Am. Chem. Soc.*, **98,** 6350 (1976).

PROBLEM 3

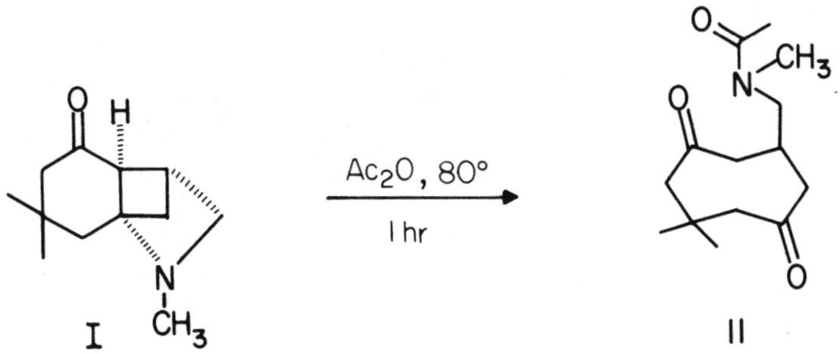

$$\text{I} \xrightarrow[\text{1 hr}]{Ac_2O,\ 80^\circ} \text{II}$$

I

II

1. Y. Tamura, H. Ishibashi, M. Hirai, Y. Kita, and M. Ikeda. *J. Org. Chem.*, **40,** 2702 (1975).

PROBLEM 3
Fragmentation of Bicyclobutane by Way of Anti-Bredt Intermediates

Some readers will be able to figure out rather rapidly at least one way to convert **I** into **II:** The lone pair of electrons on nitrogen promoting the rupture of the strained ring fusion with the aid of the electron attracting effect of the carbonyl; the ensuing iminium ion becoming a ketone with concomitant *N*-acetylation–fragmentation, and that is it. A more careful examination of this sequence, however, reveals a number of difficult issues.

First, tribute to the meticulous spectral interpretation that confirmed the structure[2] of compound **I** should be paid by looking at its three-dimensional representation (**I′**) (see Scheme 3.1). It allows one to see that this molecule embodies considerable ring strain. Disconnection of the central bridge might

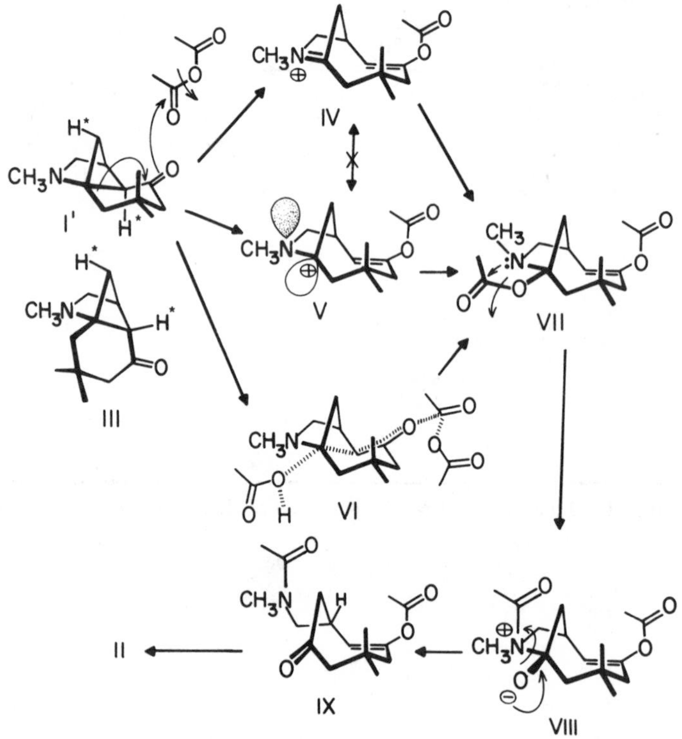

SCHEME 3.1

take place either with the assistance of the lone pair of electrons of the hetero-atom as postulated previously or solely by vibrational energy gained through collisions in the hot reaction medium. The former entails the participation of structure **IV,** whereas the latter leads to **V.** Yet, both structures are unsettling. While **IV** is clearly an anti-Bredt intermediate (see also Problem 55), the carbenium ion not only ought to accommodate itself to a nonplanar arrangement, but in addition it cannot be stabilized by the resonance effects of the vicinal nitrogen atom owing to the nearly orthogonal geometry of the two atomic orbitals involved.

If these roadblocks are overcome, **IV** and **V** would then converge to the acetate **VII,** which in turn would undergo acyl migration to the basic nitrogen (see Problem 1). The instability associated with the positively charged nitrogen in **VIII** would then provide the driving force for the final C–N bond fragmentation and the simultaneous formation of the second keto group in **II,** which would result from the hydrolysis of the acid-sensitive enol acetate during work-up operations.

Another possibility may be considered if merely a catalytic amount of acetic acid is present in the acetic anhydride medium, something to be expected in fact in any bottle of this reagent that has been opened to the atmosphere just once. If so, one would have the two molecules required for a push–pull trimolecular process portrayed in **VI** that would yield the desired product by way of a concerted unraveling reaction. However, such trimolecular couplings require a very large transition entropy, making them unlikely to the point of being experimentally unknown.

Some supporting facts for the participation of acetic anhydride may be derived from the interaction of **I**—or **III**—with weakly nucleophilic water. Interestingly, not only is the homolog of **II,** namely, **X,** produced in this fashion, but two additional components, **XI** and **XII,** have been shown to coexist with it as well.[1,3] If the formation and/or survival of α-aminohydrin **XI** is surprising, the anti-Bredt structure **XII,** which is in fact the hydroxide salt of intermediate

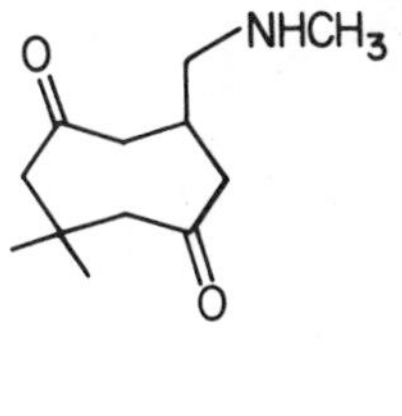

X 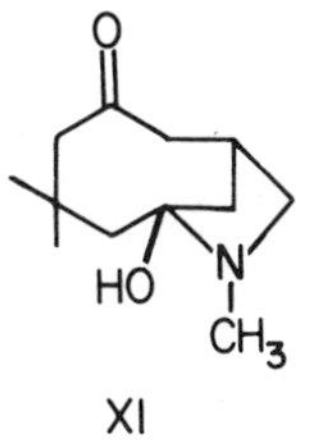 XI 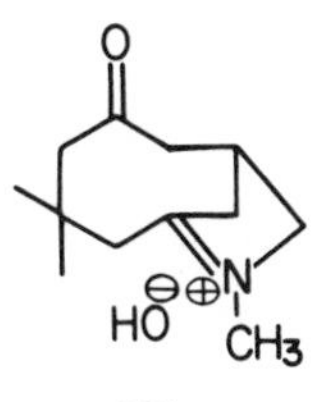 XII

IV, is a serious challenge to the unprepared mind. And yet, they have been proved to exist.

REFERENCES

2. Notably, the starred protons in **I′** show a significantly large long-range coupling (8.5 Hz) owing to the *W* configuration of this exo structure. Confirmatory evidence was collected from the lack of such coupling in the endo isomer **III.** See also J. Meinwald and A. Lewis, *J. Am. Chem. Soc.*, **83,** 2769 (1961); K. B. Wiberg et al., *J. Am. Chem. Soc.*, **84,** 1594 (1962).
3. O. Cervinka, in *Enamines, Synthesis, Structure, and Reactions*, A. G. Cook, Ed., Dekker, New York, 1969, p. 270.

PROBLEM 4

COOEt

I $\xrightarrow[\text{THF, 20°, 25 min}]{\text{LiH, } CH_2O}$ $\xrightarrow[CH_2Cl_2\text{, 20°, 20 min}]{KHCO_3\text{ *}}$ II (83%)

*Saturated aqueous solution; a two-phase system.

1. G. M. Ksander, J. E. McMurry, and M. Johnson, *J. Org. Chem.*, **42,** 1180 (1977).

PROBLEM 4
Acidic Fragmentation in the Course of a Clever Synthesis of α-Methylenebutyrolactones

This very useful reaction resulted from a failed attempt to introduce a new protected methyl vinyl ketone equivalent for the all important Robinson annulation reaction.[2] This failure turned out to be quite rewarding, for it allowed for a fast, simple, mild, and high yielding construction of α-alkylidene ketones, esters, and nitriles, as well as β-methylenebutyrolactones, a family of compounds with several representatives among the tumor growth inhibitors. The method, unfortunately, has its own limitations, which will be discussed after exposing probable mechanisms.

Compound **I** features three carbonyl centers of various degrees of reactivity. It also displays a potentially acidic proton on the carbon between the β-keto ester and lactone carbonyls. In principle, the carbanion derived therefrom should be comparable to the useful β-keto ester anions in nucleophilic substitutions and additions that lead to the formation of C–C bonds. Diethyl oxaloacetate, however, is a very poor nucleophile, because epoxides and enones remain unaltered in its presence. This peculiarity poses severe restrictions to the synthetic applications of this reaction.

In any event, the corresponding carbanion (**III**)—formed by proton abstrac-

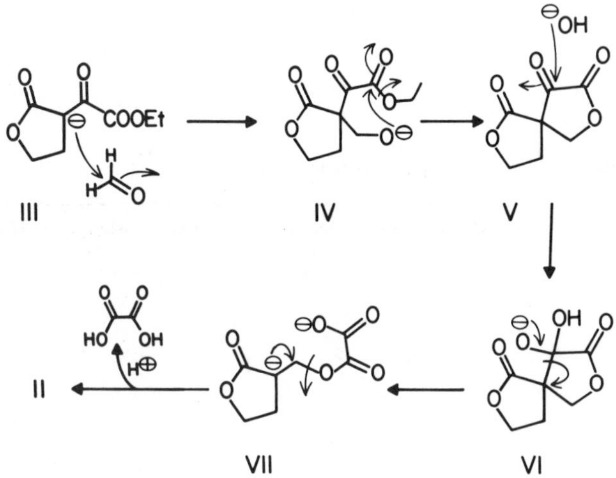

SCHEME 4.1

tion from **I** by the lithium base—is still capable of undergoing aldol condensation with the simplest of aldehydes, formaldehyde, to give a transient alkoxide that is positioned γ with respect to the highly reactive ester. This ester unit owes its increased electrophilicity to its carbonyl neighbor. As a result, an α-ketobutyrolactone ring is constructed to yield **V** (see Scheme 4.1).

In the experimental sequence, the authors at this point replace methylene chloride by THF (tetrahydrofuran) and the mixture is exposed to a mild base in a two-phase system. The three carbonyl groups of **V** should all be reactive towards nucleophilic bases such as hydroxide ion, but they also differ remarkably in their susceptibility. Of the three, the keto function would be expected to be more prone to react as a consequence of dipole–dipole interactions with its vicinal ester grouping. The intermediate ketone hydrate anion **VI** may in principle trigger the disconnection of the three bonds around the central carbon as follows:

1. The C–O bond breakage would simply reverse the process.
2. The loss of the C—C=O carbon-to-carbon bond would leave a very unstable carbonyl anion, something that has required the development of polarity inversion operations (umpolung) to be experimentally feasible.[3]
3. The remaining bond rupture would afford a relatively more stable carbanion on the butyrolactone ring of **VII,** a fundamentally acidic position widely employed in organic synthesis. The freed pair of electrons would then proceed to force the extrusion of an oxalic acid residue yielding the desired α-methylenebutyrolactone. The concerted addition–elimination process from **VI** to **II** is also a feasible alternative.

The attack of base represented in **V** is particularly instructive since it illustrates another limitation of the synthetic method: The addition–fragmentation reaction is not as clearly defined when carbonyl groupings other than γ-lactones are present. For instance, the fragmentation of **VIII** with sodium ethoxide does not furnish the corresponding enone. Instead, it yields ethyl benzoate.

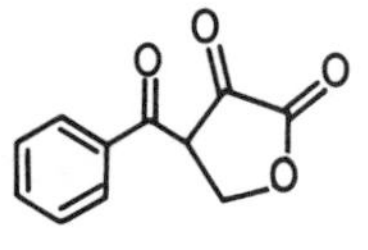

VIII

REFERENCES

2. A most successful derivative, 2-trimethylsilyl-1-propene-3-one, has been developed and is a must in the stock of reagents of any organic synthesis laboratory. See G. Stork and B. Ganem, *J. Am. Chem. Soc.*, **95,** 6152 (1973). For a convenient preparation see R. K. Boeckman, Jr., D. M. Blum, B. Ganem, and N. Halvey, *Org. Synth.*, **58,** 152 (1978). For a review on the Robinson cyclization reaction see, among other contributions, R. E. Gawley, *Synthesis*, 777 (1976).
3. B. T. Grobel and D. Seebach, *Synthesis*, 357 (1977).

PROBLEM 5

$$\mathbf{I} \xrightarrow[\text{DMF}]{Br_2} \xrightarrow{K_2CO_3} \mathbf{II}$$

COOH

I II

1. S. C. Welch, C. P. Hagan, D. H. White, W. P. Fleming, and J. W. Trotter, *J. Am. Chem. Soc.*, **99,** 549 (1977).

PROBLEM 5
A Local Oxidation at the Unexpected Site by Bromination at the Expected Site

Structures **I** and **II** differ only in the lactonic C–O bond. This seemingly simple transformation conceals complications derived from the oxidative process involved in the formation of this bond. Necessarily, the oxidant must be an external agent since the enone **II** does not show the traces (reduced fragments) that are always produced during any intramolecular redox operation. This oxidizing agent in all probability is bromine. In fact, one could predict, in the absence of any further evidence, that the reaction mechanism must proceed by way of an allylic bromination at the γ carbon of enone **I,** which would provide the required functionalization for an intramolecular nucleophilic displacement of bromide by the carboxylate anion in a second step. This is shown in structure

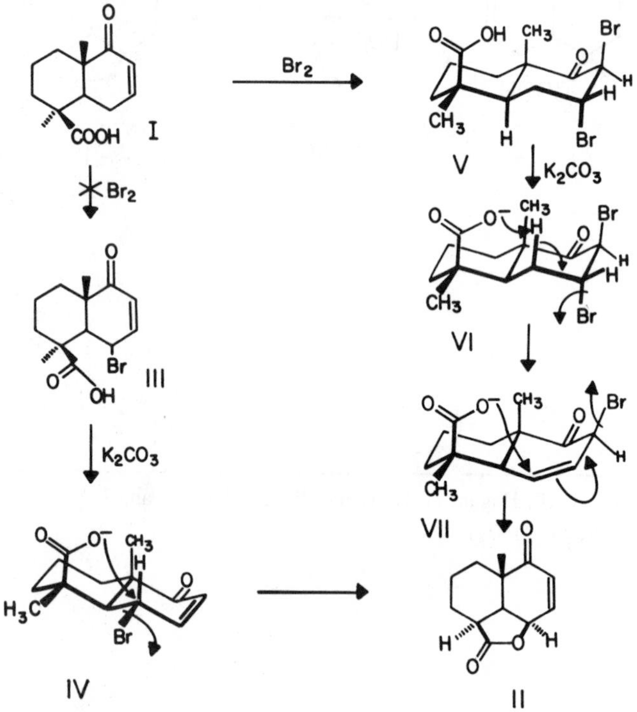

SCHEME 5.1

IV and would occur under the auspices of mild base and a conveniently polar medium such as dimethylformamide (DMF) (see Scheme 5.1).

This would be equivalent to the contention that the earth is flat because nobody had gathered evidence against this philosophically satisfactory shape. There are, indeed, some pieces of evidence against our first attempt to explain this seemingly simple reaction. The first is the consumption of one full mole of bromine in the reaction. The second is the absence of hydrobromic acid among the products of the first step. These facts can only be accommodated by addition of molecular bromine across the double bond of **I** to give a trans diaxial dibromo derivative (**V**). In actual fact, a compound featuring no vinyl protons after treatment with bromine in DMF was observed.

Exposure of this intermediate to potassium carbonate would first generate a carboxylate anion (**VI**), which in turn would be strategically located to interact with the very proximal axial proton, thus providing anchimeric assistance to the anti-dehydrobromination that ensues. The final consolidation of the lactone ring may be achieved by a S_N2' process that would cause the migration of the C=C bond to its final location. *Moral:* Never underestimate the complexity of any reaction mechanism. Often, its intricacy will be directly proportional to the available experimental data.

PROBLEM 6

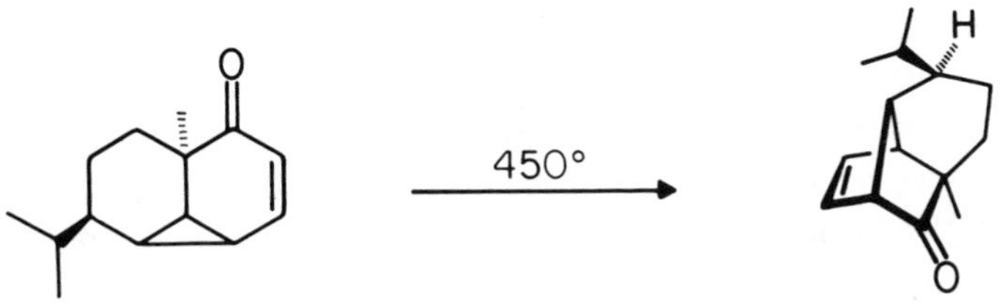

450°

I

II

---

1. J. E. McMurry and M. G. Silvestri, *J. Org. Chem.*, **41,** 3953 (1976).

PROBLEM 6
An Adorned Vinylcyclopropane Rearrangement

The solution to this problem is quite easy to visualize if the tricyclic structure **II** is looked upon as a *cis*-decalin skeleton to which an extra C–C bond between carbons in different rings has been added. That this bond is made at the expense of the cyclopropane ring of **I** becomes obvious not only because of the absence of this three membered unit in **II,** but also because the other functions (C=C and C=O) appear in the product without any change other than the isomerization of the double bond.

Conversely, the cyclopropane ring in **I** owes its reactivity to its being a vinylogous cyclopropyl ketone. This carbonyl group imbues the cyclopropane with a predictable polarization. The three-dimensional representation of the starting material shown in Scheme 6.1 allows one to see that the cyclopropane portion maintains a *bisected* geometry with respect to the vicinal double bond. This condition provides maximum overlap between the Walsh orbitals of the cyclopropane, which have been named banana orbitals, and the π molecular orbital of the enone system. This is translated in macroscopic terms as a maximum transmission of polar effects from the carbonyl down the carbon chain. As a consequence, this geometry is ideally suited for the occurrence of the well-known vinylcyclopropane to cyclopentene rearrangement[2]—actually a 1,3-sigmatropic rearrangement. A careful inspection of product **II** will reveal the presence of a cyclopentene ring as part of the bicyclic structure.

With these elements in hand a possible mechanistic sequence comes to mind: An initial regioselective cyclopropane C–C bond disconnection leads to zwitterion **III** in which the negative charge appears strongly stabilized as it

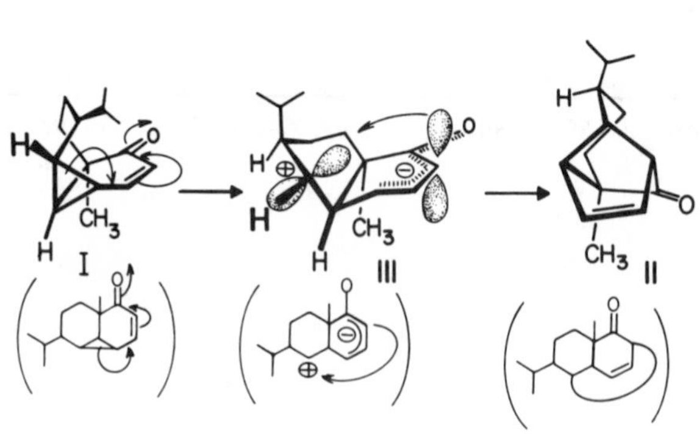

SCHEME 6.1

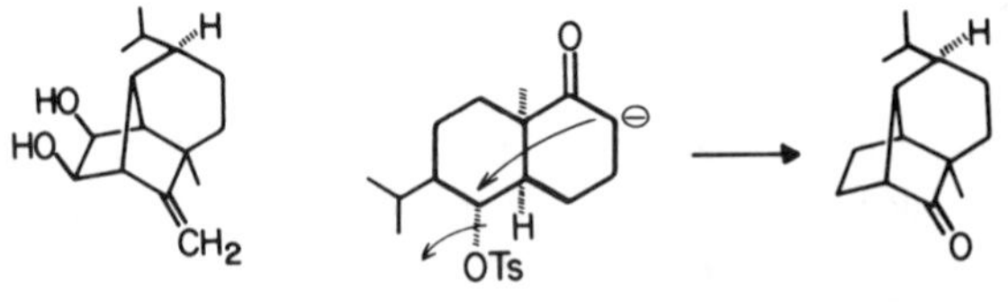

IV V VI

becomes part of the enone π orbital. This canonical structure (**III**) would contain the two opposing charges nearly at bonding distance (only the two interacting orbitals are depicted) as a consequence of the cis-ring fusion. The consolidation of this C–C bond finishes the sequence.

A homolytic cyclopropane C–C bond rupture that would furnish in turn a 1,3-diradical is also conceivable. However, it is always difficult to establish whether a purely diradical or ionic mechanism is in operation. Between these two extremes there exists a graded continuum of *polarized diradicals* of which the zwitterion represents the end of the spectrum.[3] In addition, the continuous development of radical character during the formation of the transition state of a homolytic bond scission, called the *continuous diradical*, has been postulated to explain the behavior of some reactions.[4] Alternatively, the contribution of a truly concerted transformation cannot be overlooked.[2]

It is worth noting that this transformation of **I** was the result of an unpredicted synthetic entry into ketone **II,** which in turn was originally conceived of as an intermediate en route to the Gibberlinlike plant growth promoter sativene diol (**IV**).

The original approach was designed to emulate the successful cyclization of anion **V** to the sativene precursor **VI,**[5] by means of base treatment of a closely related enone derivative. The resulting anion (**VIIa**) from the enone, however, failed to cyclize as predicted, yielding only the cyclopropane derivative **I** (see Scheme 6.2). The end result was, nevertheless, equally rewarding, since structure **III** exactly filled the purpose for which anion **VIIb** was intended.

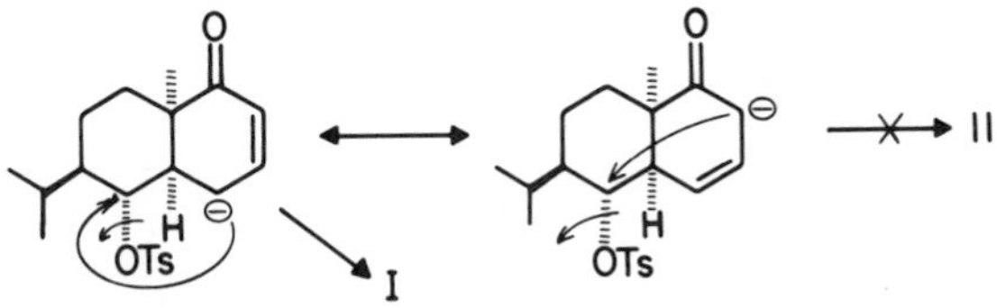

VIIa VIIb

SCHEME 6.2

REFERENCES

2. G. D. Andrews and J. E. Baldwin, *J. Am. Chem. Soc.*, **98,** 6705 and 6706 (1976) and references cited therein. See also, R. L. Danheiser, C. Martinez-Davila, and J. M. Morin, *J. Org. Chem.*, **45,** 1340 (1980).
3. J. E. Gajewski, R. J. Weber, R. Braun, M. L. Manion, and B. Hymen, *J. Am. Chem. Soc.*, **99,** 816 (1977).
4. W. von E. Doering and K. Sachdev, *J. Am. Chem. Soc.*, **97,** 5512 (1975).
5. This synthetic scheme was designed by Professor McMurry in the late 1960s. For an illuminating sequence see J. E. McMurry *J. Am. Chem. Soc.*, **90,** 6821 (1969). For a homologous, closely related case see E. Piers, R. W. Britton, and W. de Waal, *J. Chem. Soc. Chem. Commum.*, 1069 (1969).

PROBLEM 7

I → (1) 2 eq CH_3AlCl_2, 0° (2) NaOH, H_2O → II (60%)

III → (1) 2 eq $EtAlCl_2$ −80° (2) ^{-}OH → IV (14%) + V (8%) + VI (73%)

1. M. Karras and B. B. Snyder, *J. Am. Chem. Soc.*, **102,** 7951 (1980); B. B. Snyder and C. P. Cartaya-Marin, *J. Org. Chem.*, **49,** 153 (1984).

PROBLEM 7
Vicinal Migrations with Stereocontrol

Two apparently independent reactions are proposed here as part of the same problem, because of the intimate mechanistic relationship that exists between them. This is derived from their common denominator, namely, the powerful and versatile alkyl aluminum chloride reagent. These compounds have been used as Lewis acid catalysts for Diels–Alder and ene reactions, with advantages over other aluminum based Lewis acids.[2] The presence of a Lewis base (carbonyl oxygen) in starting compounds **I** and **III,** and the formation of a C–C bond in all the annullation processes that lead to cyclopentanes—something that is likely to occur by some sort of intramolecular interaction of olefin and carbonyl—clearly suggest a strategy for approaching this double sided problem.

The mechanism begins with bonding of alkyl aluminum chloride and starting materials at the carbonyl oxygen during the first stages of the sequence. This is illustrated by structure **VII,** where the keto function is converted into a delocalized cation (actually a carbonyl ylide by definition) (see Scheme 7.1). The ensuing attack by the isobutenyl group is reminiscent of the classical Prins alkene-to-carbonyl addition reaction that is assisted by protic acid.[3] A tertiary carbenium ion (**VIII**) results whose obvious end product should be Hofmann and/or Saytzeff elimination compounds (**X**). However, these were not observed experimentally.

Comparison of the aluminum alkoxide anion **VIII** and the cyclopentanone **II** shows as the principal difference a methyl group on the quaternary α carbon. The positioning of this methyl is indicative of a migration process preceding it, that is, a 1,2-hydride shift followed by, or accompanied by, synchronous 1,2-methyl migration. This highly energetic step would be triggered by the need for stabilization of the positively charged ion of **VIII** on one hand, and the instability resulting from the loss of carbonyl ylide character on the other. This lability is usually translated into strong electron donation towards the rest of the organic fragment. The combination of these two forces prompts the double migration just mentioned in a push–pull fashion, as indicated in **IX.** CD steroid ring fragment **XI** and perhydroazulene models **XII** have been synthesized conveniently by this method.

The stereoselectivity of this first reaction, illustrated by the cis relationship of the two methyls on α carbons (with respect to the ketone), deserves further comment. Assuming a degree of concert in the step between **VII** and **VIII,** one can build a molecular model with a maximum overlap of the two interacting π

SCHEME 7.1

systems as is customary in $[\pi + \pi]$ cycloadditions, by twisting this model in a comfortable conformation such as that depicted in **XIII.** This arrangement is possible only if the methyl on the optically active carbon is on the opposite side of the alkene portion so it is able to come sufficiently close to the carbonyl fragment. A few moments of rotating the model will convince us that it is impossible to achieve this approach otherwise. The overlap of the two shaded π orbitals in **XIII** that yields cyclization forces the methyl of the acyl group upward, while at the same time moving the isopropyl group downward (two wide arrows in the figure attempt to show this relative motion). As a consequence, the two vicinal methyls end up in a cis configuration.

Furthermore, if it is true that the 1,2 shift leaves a planar carbenium ion **IX,** a process that erases the configurational identity of the carbon atom bearing the isopropyl group (and therefore apparently any further use of the advantages first introduced by the stereoselective step **VII** → **VIII**), it is also true that the ensuing 1,2-methyl migration (**IX** → **II**) must proceed in a suprafacial fashion

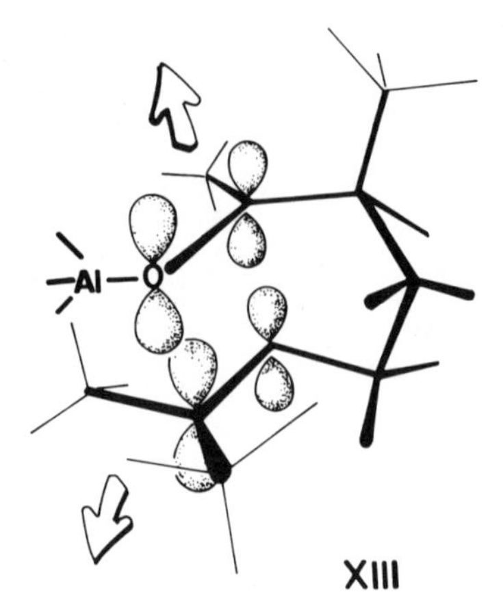

in order to follow symmetry rules; namely, the migrating methyl remains on the same side of the ring. The consequence of all this is the cis configuration observed in product **II.**

As for the cyclization of **III** in the second part of this problem, homologation of these arguments is insufficient to secure a reasonable mechanistic proposal, since new elements are introduced. First, only cyclopentanone (**V**) can be explained by a mechanism similar to that exposed here, which the reader should be able to work out without further comment. Second, an ethyl group is incorporated in compound **IV.** Third, ethylene is produced, indicating, therefore, that a reductive step is in operation over the organic portion. In fact, end products **IV** and **VI** are cyclopentanols, the result of reductive annullations.

This means two things: (1) the final 1,2-hydrogen shift that secures the formation of the ketone on the ring does not occur in the sequence that yields **IV** and **VI,** a circumstance that carries the implication of more favorable pathways for the decomposition of the aluminum alkoxide anion. (2) If one reduction takes place, one should look for one transferable hydride to the carbenium ion **XV** that should result in turn from the cyclization step analogous to that suggested in **VII** → **VIII.** This hydride is likely to be supplied by the ethyl group on the aluminum atom in the way indicated in **XV** (route a, Scheme 7.2), since it would also yield the observed ethylene molecule and would bring stabilization to the aluminum alkoxide anion by taking away excess electrons. This is in essence a peculiar case of a normal β elimination.

This transfer would possibly take place through the eight-membered, configurationally favored transition state depicted in **XVII.** Such a highly ordered

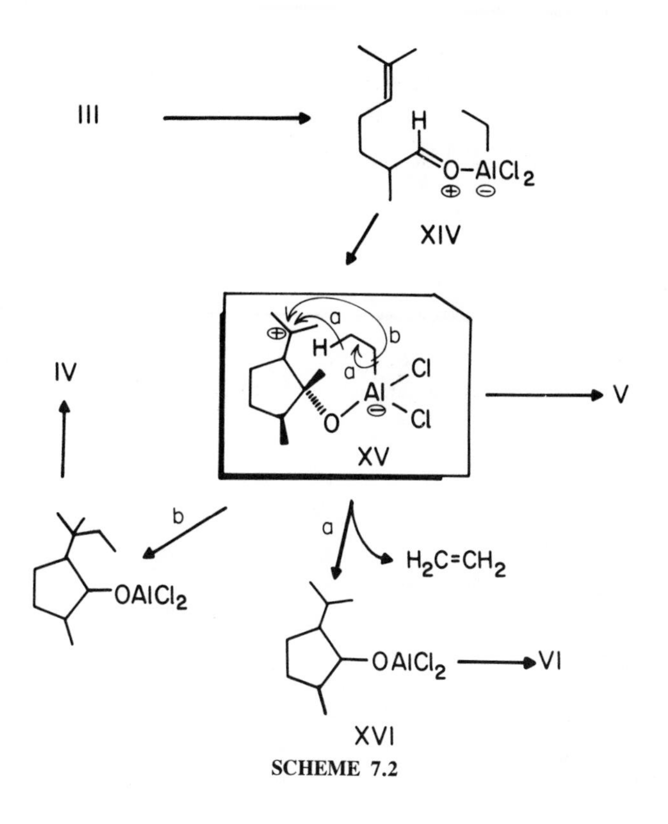

SCHEME 7.2

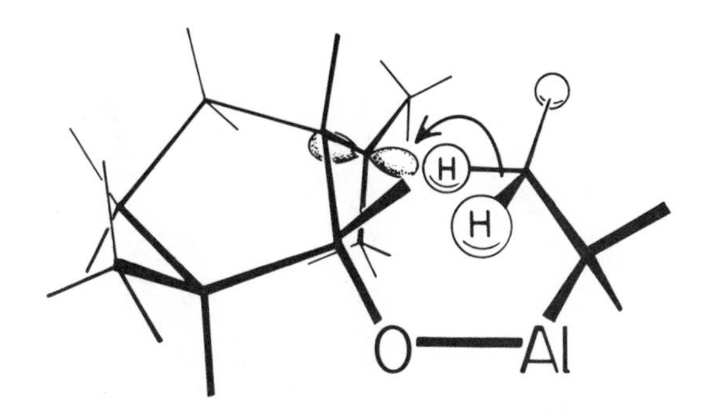

XVII

transition state implies a negative entropy change that is in fact consistent with the observed increase in the proportion of **VI** at lower temperature (73% at −80°C, 41% at 0°C). Finally, a similar transfer of the entire ethyl substituent to the central carbonium ion (**XV**) (route b, Scheme 7.2) would furnish compound **IV**.

REFERENCES

2. For a convenient review see B. B. Snyder, *Acc. Chem. Res.*, **13,** 426 (1980).
3. D. R. Adams and S. P. Bhatnagar, *Synthesis*, 661 (1977).

PROBLEM 8

N–NH–S(=O)(=O)– + B-H

I II

NaOAc
$CHCl_3$

III

---

1. G. W. Kabalka, D. T. C. Yang, and J. D. Baker, *J. Org. Chem.*, **41,** 574 (1976).

PROBLEM 8
Molecular Nitrogen and the Decarbonylation of an Enone

This reaction involves a curious reduction. The extent of this reduction—and hence an important element of judgment for devising a mechanism—becomes apparent by comparison of the ethyl-cyclohexenyl fragment of starting material and product. While **I** is in the oxidation level of an enone, the product is a monoalkene. This means, in the absence of further bond reorganization, there are two reductive steps involved. Here one encounters the first difficulty: Only one equivalent of the presumed reducing agent, catechol borane—a junior member of the extended and incredibly useful family of boron hydrides—was used. This means that only one hydride was transferred. Consequently the second reductive step must entail the oxidation of some separable component of the starting compound. In addition, this internal redox process must occur in such a way as to move the double bond to the thermodynamically unfavored exocyclic position, which, by the way, has been a synthetic difficulty over the years.

Now, everyone with a rudimentrary background on metal hydride reductions knows that a formal addition of the elements M–H to the C=O bond takes place at a first stage. There is no reason to believe that anything different occurs to a close relative of carbonyls such as the hydrazone group upon exposure to boron hydride. On that account, the postulation of structure **IV** is as reasonable as the metal alkoxides that have always been proposed for analogous reductions of aldehydes and ketones (see Scheme 8.1).

Adduct **IV,** in turn, features a boron atom and an electron withdrawing substituent (tosylate) on vicinal nitrogen atoms of sp^3 hybridization, an arrangement that is very reminiscent of a R–B–C–C–X system (X = OR, NR, SR), which is known to undergo β elimination under very mild conditions.[2] The end result is a diazene intermediate (**V**) that is ideally suited to bring about the

I + II → IV → V → III

IV: H, N, N, H, Ts, B, O, O, ⊖OAc

V: N, N, H

BOAc + TsNa + N≡N

SCHEME 8.1

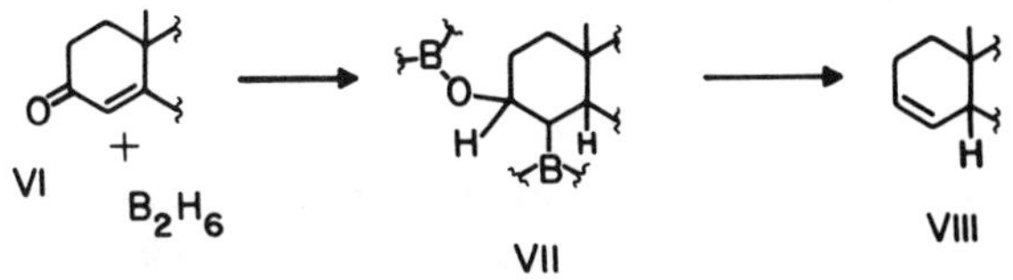

VI + B_2H_6 → VII → VIII

SCHEME 8.2

desired oxidized species: that is, molecular nitrogen! This might be accomplished either by intervention of sodium acetate, which acts as a base to remove that proton on nitrogen, or by means of a concerted 1,5-hydrogen transfer as shown in **V.** This would cause the aforementioned selective, unfavorable migration of the double bond.

This sequence of events is in contrast with the reduction of α,β-unsaturated ketones with diborane, a process that is known to start with the addition of diborane to the C=C and C=O bonds that yields **VII** (see Scheme 8.2).[3] This intermediate subsequently undergoes elimination to the alkene **VIII** wherein the C=C linkage appears shifted. (*Remember:* **VI** requires two equivalents of BH_3 for its conversion to **VIII,** since the ketone does not contain the strategic components for its self-redox as compound **I** does.)

The use of sodium acetate·trideuteriohydrate causes the selective incorporation of deuterium at the β carbon of the original unsaturated tosyl hydrazone. This datum alone, however, does not warrant a clear distinction between the various possibilities open to this system, as Scheme 8.3 shows.

In general, the reduction of tosyl hydrazones with boron hydrides[4] is a convenient alternative to the classical procedures of Wolff–Kishner, Clemmensen, and Raney nickel reductions[5] when one wants to delete the carbonyl of ketones and aldehydes after using it for the very many synthetic operations it allows.

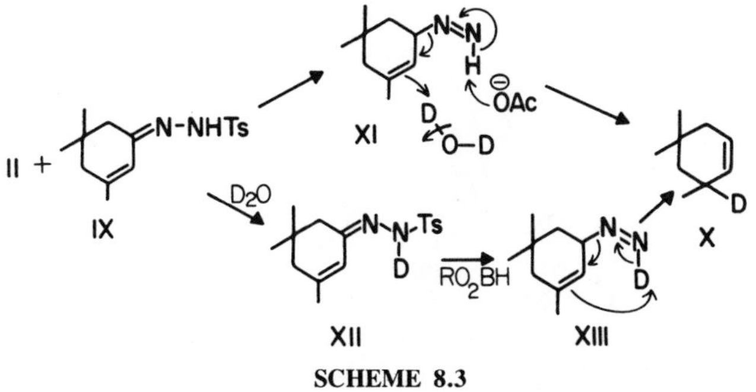

II + IX (N-NHTs) → XI (N=N–H, ⊖OAc, D–O–D) → X; IX + D_2O → XII (N–N(D)–Ts) → RO_2BH → XIII (N=N–D) → X

SCHEME 8.3

REFERENCES

2. D. J. Pasto and R. Snyder, *J. Org. Chem.*, **31,** 2777 (1966).
3. L. Caglioti, G. Canielli, G. Maina, and A. Selva, *Tetrahedron*, **20,** 957 (1964).
4. R. O. Hutchins, M. Kacher, and L. Rua, *J. Org. Chem.*, **40,** 923 (1975).
5. See, for example, L. Caglioti, *Tetrahedron*, **22,** 487 (1966).

PROBLEM 9

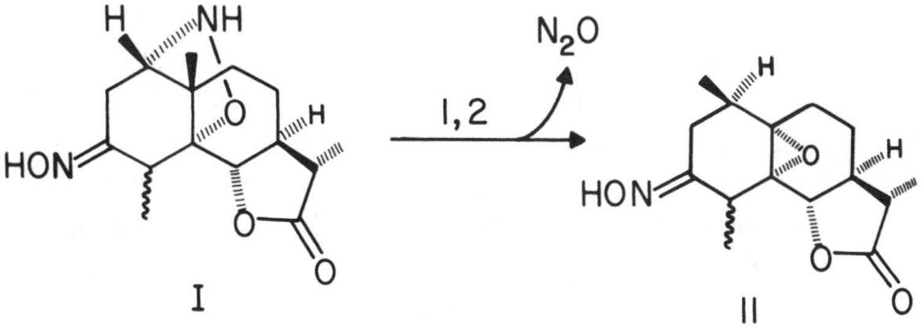

1 = $NaNO_2$, HCl (aq), 20°. 2 = HOAc, H_2O, Δ

1. J. T. Edward and M. J. Davis, *J. Org. Chem.*, **43,** 536 (1978).

PROBLEM 9
1,2-Alkyl Migration with Double Inversion of Configuration

For those who are familiar with natural products chemistry it will be easy to recognize in structure **I** the main features of α-Santonin (**III**). This was a heavily investigated compound during the 1950s and 1960s, especially its photochemistry.[2] The particular functionality that attracted so much attention is the cross-conjugated enone moiety. It is this function in fact that warranted the construction of the starting compound **I** by way of a double Michael-type addition of hydroxylamine.[3]

Product **II** may contain sufficient information in its structure to provide leads to at least one feasible mechanism. First, it displays a C–O bond, now part of an oxirane, at the tertiary angular carbon where the angular methyl of **I** used to be. At the same time, this displaced methyl appears on that carbon atom where the nitrogen in **I** was located (C-1). This atom and group reorganization seems to follow the well established lines of a 1,2-alkyl migration, which in turn requires at some point the development of a low electron density center, evidently on the C-1 carbon.

Second, that a new N–N bond is being formed in the process is suggested by the production of nitrous oxide, something that provides a convenient escape route to the amino group on C-1. This nitrosation must precede the construction of the epoxide, a task that should be fulfilled by nitrous acid.

Therefore, intermediate **IV** may be safely postulated (see Scheme 9.1). When subjected to strong protic acid (HCl) the *N*-nitroso group of **IV** activates the C–N bond by providing an electron sink, which is a desirable component in the formation of carbenium ions. It is this carbocation that allows the Wagner–Meerwein rearrangement of the angular methyl that is illustrated in **V**. The more stable tertiary carbenium ion of **VI** is then ideally set up for the epoxide ring closure, following the departure of nitrous oxide from **VI**.

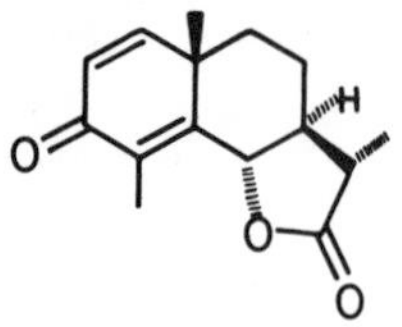

O H O O

III

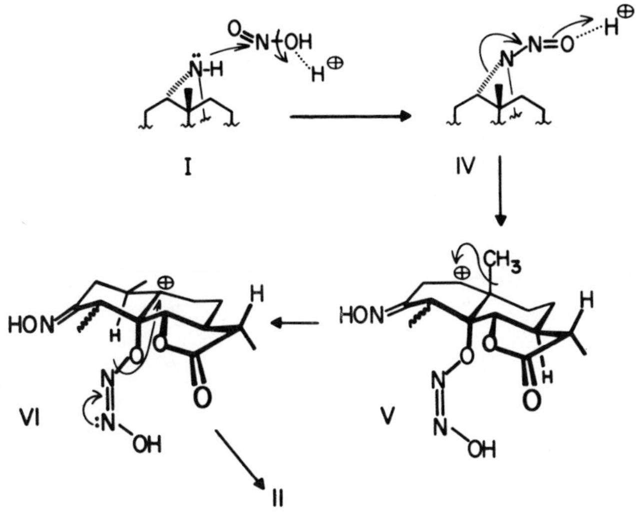

SCHEME 9.1

REFERENCES

2. For the early story of Santonin and its informative chemistry see J. L. Simonsen, and D. H. R. Barton, *The Terpenes*, Vol. III, Cambridge University Press, New York, 1952, p. 292; P. J. Kropp, *Organic Photochemistry*, O. L. Chapman, Ed., Dekker, New York, 1967, pp. 1–86.
3. This reaction was first explored many years ago, but the proposed structures of observed products required substantial modification. This is one case where chemical degradation could not provide all the structural answers. See L. Francesconi and G. Cusmano, *Gazz. Chim. Ital.*, **39(II),** 121 (1909).

PROBLEM 10

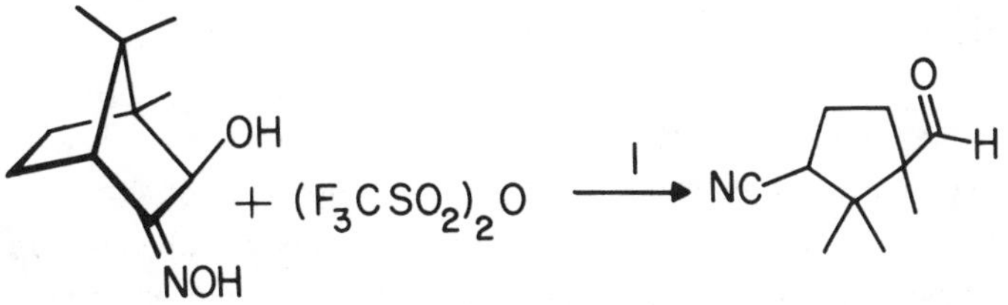

I + $(F_3CSO_2)_2O$ —I→ II (65–81%)

I = Et_3N, CH_2Cl_2, 20–50°, 4hr.

1. G. A. Olah, Y. D. Vankar, and A. L. Berrier, *Synthesis*, 45 (1980).

PROBLEM 10
A Facile Fragmentation with a Difficult Electron Budget

The detailed treatment of this rather simple problem leads to some concepts that should be clear to anyone who wants to solve mechanistic problems intelligently and conscientiously.

If one considers the general principle that the heterolytic breakage of a C–C bond results in the relative oxidation of one carbon end (the carbocation) and the simultaneous reduction of the other carbon end (the carbanion), or groups directly attached to these atoms, it is possible to infer that the nitrile group of **II** finds its antecedent in the oxime function and the aldehyde in the carbinol group of **I,** if one allows for the heterolytic rupture of the C–C bond that separates them.

However, in this case, while it is obvious that carbinol yields aldehyde by a one-step oxidation contained in the C–C bond rupture, it is somewhat harder to visualize the additional oxidation (not reduction, as the direction of the C–C bond breakage suggests) of the oximino group that should be responsible for the construction of the final nitrile. Oxidation of one group means reduction of another in a chemically balanced equation. Hence one should look around for a reduced fragment. This demands particular attention to all negatively charged species or potential precursors. Among these, those containing strong electron withdrawing substituents appear as suspects. In this problem, the most likely candidate is the trifluoromethyl sulfonyl function (triflate) of the trifluoromethylsulfonyl anhydride. Moreover, its anhydride character imbues the sulfur atom with a strong electrophilic nature. It is therefore safe to propose the attack of the oxygen atom of the oxime function on the sulfur with displacement of a stable triflate anion (the negatively charged species we were seeking).

Subsequently, triethylamine would abstract the only acidic proton of the organic fragment (the alcohol) as indicated in **III** (see Scheme 10.1). The liberation of O–H bonding electrons and the ease of departure of the triflate anion (the best leaving group ever) induce the rupture of the carbon framework. It is this triflate, the anion, that carries away the excess electrons of the carbon skeleton, leaving behind an oxidized species, which is the observed nitrile.

This cleavage reaction can be performed with a host of reagents[2]: phosphorous pentachloride, benzenesulfonyl chloride, phosphoric acid, thionyl chloride, phosphonitrile dichloride, and dichlorocarbene.[3] All these reactions, nevertheless, use conceptually the very same construction of an ester followed

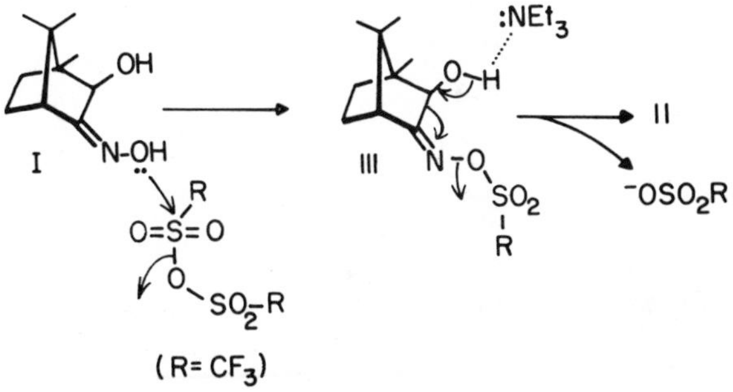

SCHEME 10.1

by fragmentation of a **III**-type intermediate, a sequence of events that is reminiscent of the Beckmann rearrangement.

REFERENCES

2. J. N. Shah, Y. P. Mehta, and G. M. Shah, *J. Org. Chem.*, **43,** 2078 (1978).
3. C. A. Grob and P. W. Schiess, *Angew. Chem. Int. Ed. Engl.*, **6,** 1 (1967).

PROBLEM 11

NaH / DMSO / 40°-50° / 1hr → H_3O^+ →

SO$_2$Ph

II

PhSO$_2$

III

I

1. D. A. Chass, D. Buddahsukh, and P. D. Magnus, *J. Org. Chem.*, **43,** 1750 (1978).

PROBLEM 11
A Sulfone Assisted Carbanion and a Tosyl Group Jointly Take Away a Carbonyl Group

Another set of bicyclic structures has been included in this problem to remind the reader that an organic chemist with a firm knowledge of conformational analysis is better prepared to face complex mechanistic situations.

This two-faceted problem requires consideration in two separate stages, since two apparently disparate mechanisms appear involved in the production of compounds **II** and **III.** For one thing, while **II** contains the same number of skeletal carbons as does the starting tosylate, one carbon, two hydrogens, and an oxygen atom are missing in the final structure (**III**).

In strong bases such as the one provided by sodium hydride and dimethyl sulfoxide (DMSO)—namely, dimsyl sodium—one should expect the formation of carbanions at sites of acidic protons. Ketones are attractive as potential sources of carbanions. However, ketone **I** features two blocked α carbons, without protons. Conversely, the tosyl group is ill suited for carbanion stabilization. The last functionality one may appeal to is the phenyl sulfone substituent at the end of the *sec*-pentyl chain. Recent investigations have revealed their potential as carbanion precursors, adding an important feature to their considerable usefulness in organic synthesis: That is, sulfones can be removed under such mild conditions that carbonyl groups are not affected, and unconstrained α-sulfonyl carbanions have the unusual quality of retaining the asymmetry of their precursors in a wide variety of experimental conditions.[2]

It seems likely as a consequence that a carbanion may be formed under these experimental conditions at the carbon vicinal to the sulfone group. This anion then would be conveniently placed at a six-carbon-atom distance from the electrophilic carbonyl, thus providing an expedient base to form the C–C bond clearly required for the construction of the six-membered ring of **II.** However, ketones become tertiary alcohols upon attack by carbon nucleophiles. Consequently, one of the alpha C–C bonds next to this ketone must be broken in order to preserve the ketone of the final product. The electron reorganization that would follow is consistent with a concomitant extrusion of tosylate anion. This is illustrated by three-dimensional structures **IV** and **V** (see Scheme 11.1)

An example of the strong electron withdrawing power of aryl sulfones is provided by the fact that compound **II** undergoes subsequent fragmentation to carboxylic acid **VI** when traces of water are present in the reaction mixture.

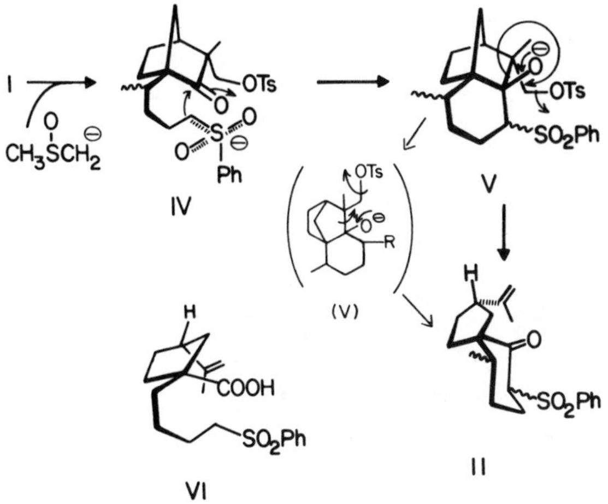

SCHEME 11.1

It is somewhat more difficult to account for product **III.** A reasonable starting point could well be intermediate **V** since it features the same tricyclic structure present in the target compound. In addition, since skeletal carbons are missing, namely, the CH_2O unit indicated previously, the mechanism must, of course, include at some point a carbon fragmentation. A handy method for the departure of a formaldehyde fragment involves the prior construction of an oxetane ring followed by a retro [2+2] cycloaddition as shown by structures **i–iii**.

The tosyl and alkoxide units of **V** would be suitable for this sequence were it not for the stereochemical impossibility of forming a four-membered ring from the two vicinal endo and exo substituents present in **V.** One might argue that there must be a mistake in the configurational assignment of the quaternary carbon (denoted C*) because the oxetane scheme would be feasible if **V** were just the other epimer. To this argument's misfortune, compound **I** as depicted was obtained unambiguously from the rather interesting Lewis acid-catalyzed deep-seated rearrangement of the tricyclic ether[3] **VIII,** which, in turn, was constructed in a stereospecific manner from (+)-nopinone (**VII**), a natural product

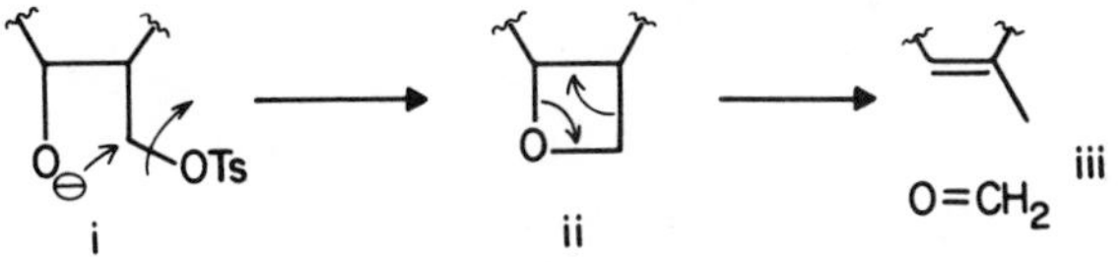

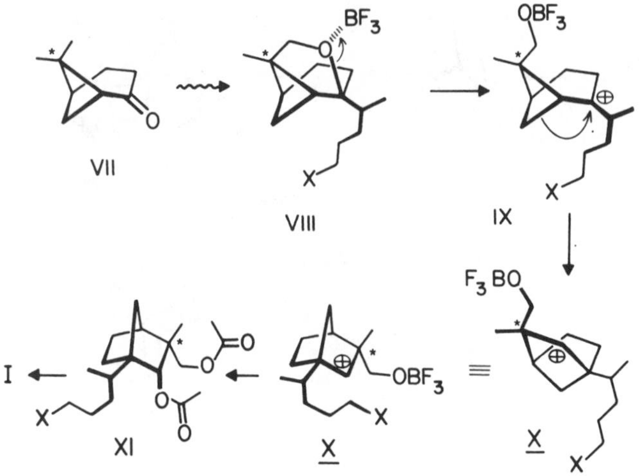

SCHEME 11.2

whose absolute stereochemistry has been well established (see Scheme 11.2). Consequently, the quaternary carbon in question (C*) must end up unmistakably in the (*R*) configuration.

One may overcome this inpasse by changing the direction of approach of the arylsulfone carbanion to the ketone in **IV** from the endo to the exo side of the molecule, as depicted in **IV′** (see Scheme 11.3). This is not only sterically feasible, but in addition the alkoxy and tosylate units would be parallel to one another at close to bonding distance. Their collapse to the oxetane postulated previously would occur easily as portrayed in **XII** → **XIII.**

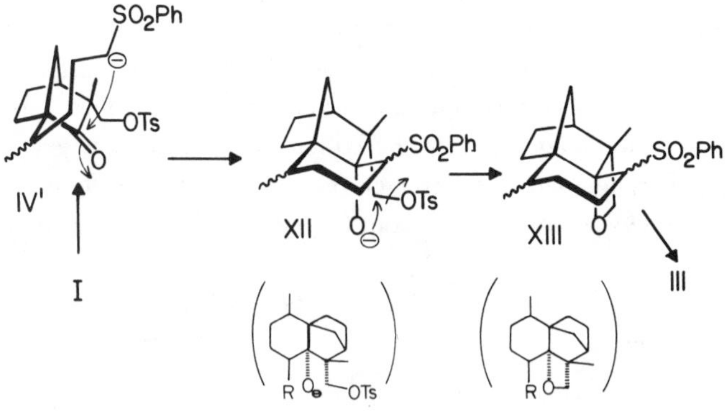

SCHEME 11.3

REFERENCES

2. D. J. Cram, W. D. Nielsen, and B. Rickborn, *J. Amer. Chem. Soc.*, **82,** 6415 (1960).
3. T. W. Gibson and W. F. Erman, *J. Am. Chem. Soc.*, **91,** 4771 (1969).

PROBLEM 12

OH

\+ $(CH_3)_2N\text{-}CH(OEt)_2$

II

I

EtOH (trace)

H^+ (trace)

$N(CH_3)_2$

$(CH_3)_2N$

\+

III (24%)

IV (40%)

1. K. A. Parker, R. W. Kosley, Jr., L. Buchwald, and J. J. Petraitis, *J. Am. Chem. Soc.*, **98,** 7104 (1976).

PROBLEM 12
Transfer of an Amino Group without Assistance from Transition Metals

The fact that the carbon backbone of the starting tertiary alcohol **I** appears with little modification in both products strongly suggests that enamines **III** and **IV** were formed by addition of the elements of dimethylamine to the acetylene end of compound **I.**

Here lies the first roadblock. Although the large scale preparation of some related vinyl derivatives such as enol ethers from acetylene itself has been used in the chemical industry for years, these syntheses usually require catalysis by mercuric salts[2] and yields are often quite poor. Similarly, enamines have been claimed to be intermediates in the synthesis of amines from alkynes, but this process also requires mercuric ion catalysis.[2]

Except for the metallic impurities present on the walls of poorly washed glassware,[3] no other transition metal catalyst may be assumed to be present in this experiment. Consequently, the transfer of the dimethylamino group from the ortho amide **II** to **I** must be taking place under less energy demanding conditions, where protic acid catalysis suffices to activate the triple carbon to carbon bond.[4]

One particularly effective way nature uses to reduce activation energies and other molecular requirements is by the intensive use of intramolecular reactions. In this case the interacting species are held together on the same molecular template at a very short distance, thus paving the way to regio- and stereospecific reactions.

Structures **I** and **II** may, in fact, form the desired intermediate in which the nitrogen-bearing function is to be transferred intramolecularly to the alkyne fragment taking advantage of the high sensitivity of ortho acid derivatives toward mild nucleophiles. The ortho ester product **IV** is a hint of the coupling of compounds **I** and **II** through the oxygen atom of the former, a process that includes the simultaneous exclusion of ethanol (see Scheme 12.1).

Intermediate **V** thus formed would allow for the transfer of the dimethylamino group to the alkyne by virtue of a push–pull effect of coordination of the triple bond with the catalytic proton and the electron releasing effect of the two ether oxygens. The discoverers of this reaction assumed that "it proceeds by way of carbonium ions (**VI** and **VII**) which subsequently eliminate or add ethanol" to give **III** and **IV**, respectively.

Alternatively, the formation of enamine derivatives may also be the con-

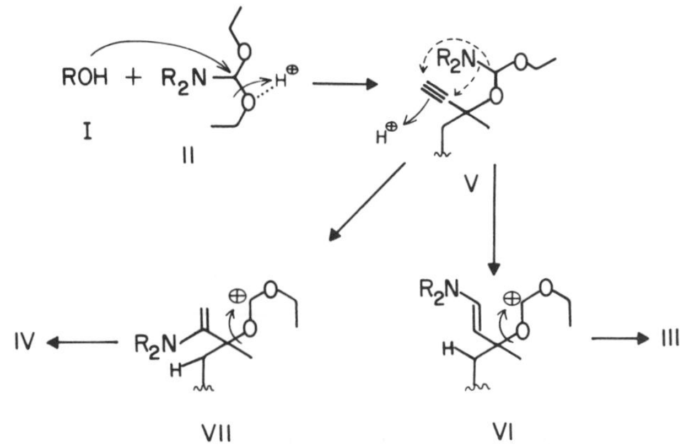

SCHEME 12.1

sequence of other diverging routes stemming from the key intermediate **V.** This structure might evolve into either a five- or six-membered transition state (illustrated by structures **VIII** and **IX** of Scheme 12.2). The dihedral angle of the three carbons of a propargyl unit is always thought of as 180°. This would constrain the approach of the nitrogen atom to this π system. However, the bending of the C–C–C framework of acetylenes requires surprisingly little energy (0.3 kcal for a 6° bend[5]). Thus, in a transition state of some 40 kcal/mol above ground state, acetylenic substrates can assume the same geometry as their olefinic analogs.[6] It is, therefore, conceivably a six-membered transition state for the transfer of the amino unit.

A concerted electrocyclic reaction such as that in **VIII** would give rise to an allene **IX** whose 1,3-hydrogen transfer would yield the final compound **III.** However, such an electrocyclic reaction does not, as yet, have precedent,[7] and therefore may not be the most probable route. In addition, a nonconcerted electron reorganization would also furnish allene **IX** via a stable allylic carbocation (**X**).

On the other hand, the bending of the acetylene unit aided by its association with the proton would allow for the approach of the amino group to the other acetylene carbon. The transition state corresponding to this idea (**XI**) would evolve into either carbocation **VI** (see Scheme 12.1) or cation **XII** depending on which C–O bond is concomitantly broken. Proton elimination would finish this sequence.

Unfortunately, the available data do not warrant a judicious, well supported choice among the various pathways.

SCHEME 12.2

REFERENCES

2. P. F. Hudrlick and A. M. Hudrlick, *J. Org. Chem.*, **38,** 4254 (1973).
3. Even careful experimenters stumble into this pitfall. For example, see reference (9) of Problems 38 and 16. Furthermore, it is surprising to learn that a number of well established reactions will not go if the reagents and glass equipment are exhaustively cleaned of metallic impurities. An example is the Grignard reaction.
4. See, for example, J. B. Lambert, J. J. Papay, and H. W. Mark, *J. Org. Chem.*, **40,** 633 (1975).
5. A. Viola, J. H. MacMillian, R. J. Proverb, and B. L. Yates, *J. Am. Chem. Soc.*, **93,** 6967 (1971).
6. H. Kwart, S. F. Sarner, and J. Slutsky, *J. Am. Chem. Soc.*, **96,** 5234 (1972).
7. For a related proposition see J. M. Reuter and R. G. Salomon, *Tetrahedron Lett.*, 3199 (1978).

PROBLEM 13

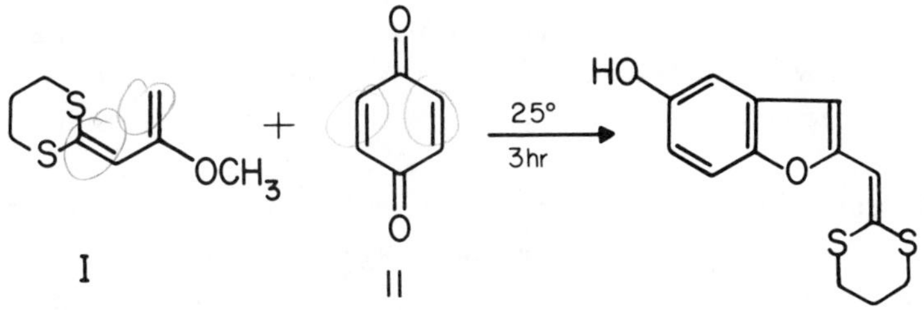

I + II $\xrightarrow[\text{3hr}]{25^\circ}$ III (62%)

1. S. Danishefsky, R. McKee, and R. K. Singh, *J. Org. Chem.*, **41,** 2933 (1976).

PROBLEM 13
Stepwise Diels–Alder Addition That Yields a [3 + 2] Adduct

If this problem consisted of proposing a likely product of the reaction of **I** and **II,** it would certainly be hard to resist the temptation to draw structure **IV** as that product, since it would result from the Diels–Alder cycloaddition of diene **I** to the powerful and classical dienophile **II.**

As a matter of fact compound **I** was designed with this particular purpose in mind, as part of a broad research program aimed at the development of highly functionalized dienes that would pave the way towards more reactive cyclohexenes than those previously obtained by means of this sort of cycloaddition.[1] The fundamental concept is illustrated in Scheme 13.1.[2,3]

To the fascination of its creators, compound **I** reacted as predicted for a normal diene with simple dienophiles such as methyl vinyl ketone, methacrolein, methyl acrylate, maleic anhydride, and tetracyanoethylene, with yields of about 30%.

Exposure of **I** to another efficient dienophile, quinone **II,** yielded the abnormal product **III.** This compound shows signs of *C*- and *O*-alkylation processes that call for the operation of a mechanism other than the normal [2+4] cycloaddition. 1,4-Benzoquinone is not only susceptible to Diels–Alder operations but also to Michael-type 1,4 additions. In this case the required nucleophile would be the vinyl ether portion of **I** while the alkyl dithiane group would remain unaltered throughout the sequence. Its role would be limited to provide substantial stabilization to the carbenium ion in **V.** The second *C*-alkylation step that would lead to four- and six-membered ring structures **VI** and **IV,** respectively, is overcome by proton transfer and *O*-alkylation to yield **III** (see Scheme 13.2).

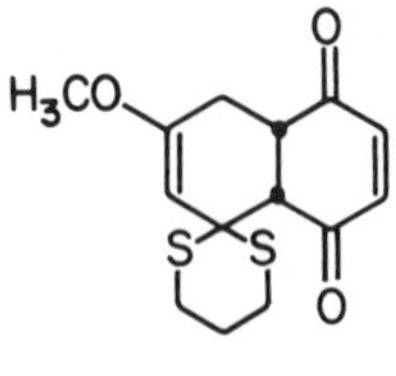

H_3CO O S S O

IV

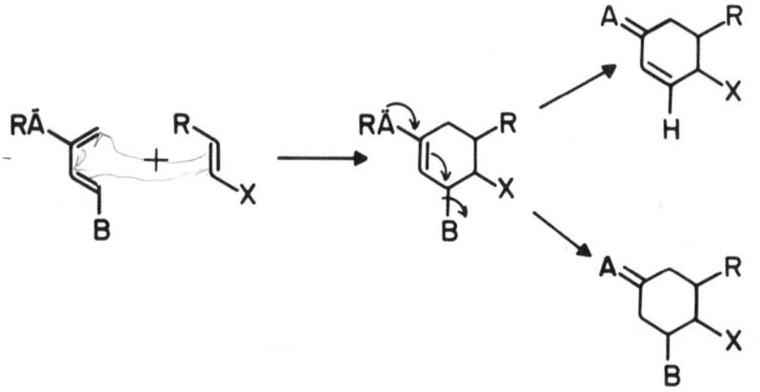

SCHEME 13.1

A strong indication that this is indeed the reaction course is provided by the construction of compounds **VIII–X** from the addition of **I** to dimethyl acetylene dicarboxylate, dimethyl azodicarboxylate, and 4-phenyl-2,4-triazoline-3,5-dione, respectively.[1]

This unexpected behavior of diene **I** was interpreted "in terms of the strongly nucleophilic character (of **I**), which is attacked by highly reactive un-

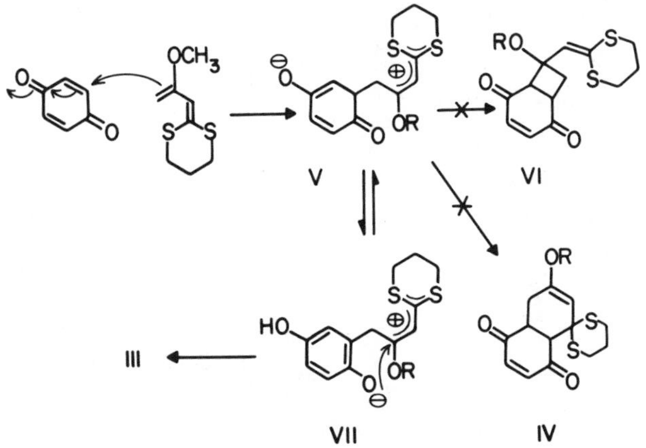

SCHEME 13.2

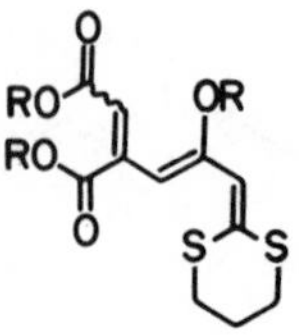

VIII 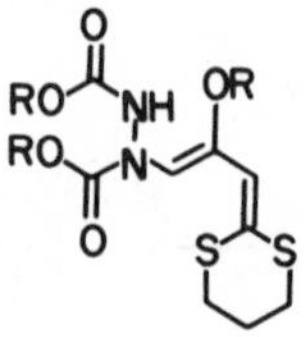 IX 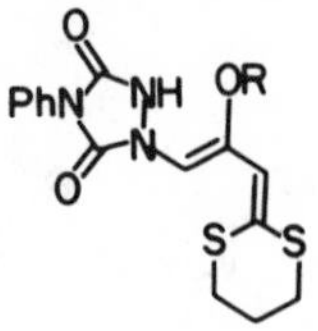 X

saturated electrophiles in orientational arrangements which are not suitable for concerted cycloaddition.[1]''

REFERENCES

2. S. Danishefsky and T. Kitahara, *J. Am. Chem. Soc.*, **96,** 7807 (1974).
3. S. Danishefsky, R. K. Singh, and R. B. Gammill, *J. Org. Chem.*, **43,** 379 (1978).

PROBLEM 14

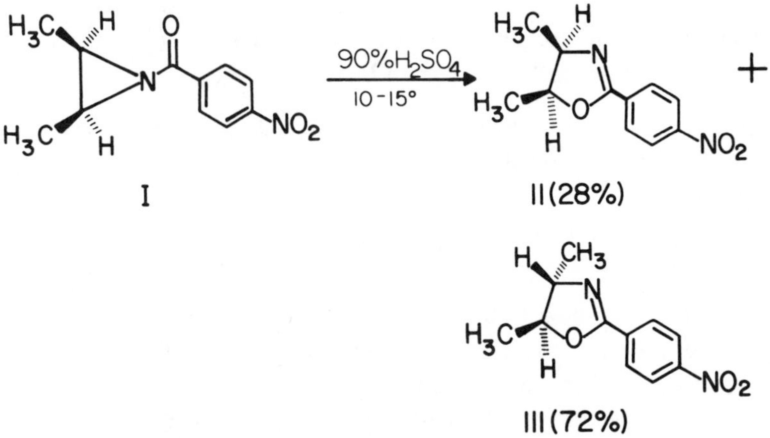

H_3C, H, O, N, NO_2 — I

$\xrightarrow[10-15°]{90\% H_2SO_4}$

II (28%) + III (72%)

1. S. P. McManus, R. A. Hearn, and C. U. Pittman, Jr., *J. Org. Chem.*, **41,** 1895 (1976).

PROBLEM 14
Transition from Concerted to Stepwise Route in an Apparent 1,3-Sigmatropic Rearrangement

The mechanism of this reaction is an example of a process whose complexity is concealed by the mirage of its apparent simplicity. The simple approach to this problem would be based on the recognition of the fact that the disconnection of the C–N bond of aziridine (**I**) followed by the formation of a C–O bond suffices to explain the conversion of **I** to oxazolines **II** and **III.**

Complications arise at the beginning because at least two feasible mechanisms that follow this sequence of events can account for the formation of oxazolines of mixed stereochemistry from **I**:

(a) Either these two bond-changing episodes take place in a concerted fashion, which has been claimed to occur in the closely related vinyl cyclopropane systems[2] or . . .

(b) the C–N bond yields first, leading to a carbocationic species that subsequently cyclizes to **II** and **III.**

The latter proposition seems more attractive for the present case if the transformation of the unsymmetrically substituted aziridine **IV** to **V**—and not **VI**—is taken into account (see Scheme 14.1).[3]

Conversely, the presumed heterolytic C–N bond separation should involve a prior protonation step since this reaction takes place only under strong acid catalysis when it is run at room temperature. In the absence of acid, however, the same process may be thermally induced. There are two potential sites for proton attachment: The aziridine nitrogen and the carbonyl oxygen. Both—as protonated species—are suitable to initiate fragmentation of the three-membered ring by way of intermediates **VII** and **VIII,** respectively (see Scheme 14.2). While evidence supporting the existence of **VIII** is available from proton nmr spectral analysis of a *N*-acrylaziridine in superacid media,[4] other researchers[1,3]

IV V VI

SCHEME 14.1

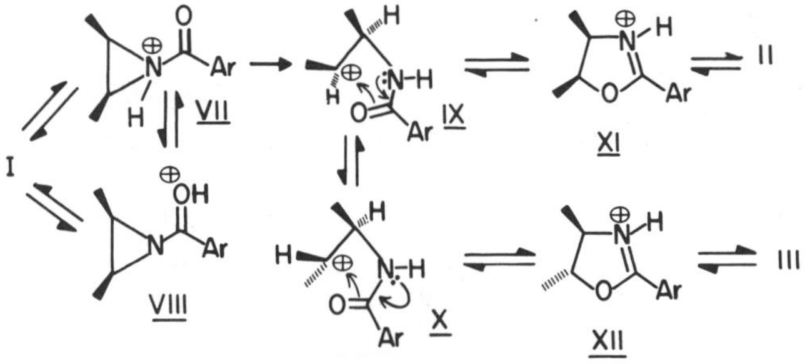

SCHEME 14.2

favor the *N*-protonated species, based on the chemical behavior of these compounds. Nevertheless, absolutes are always difficult to establish, and the likelihood of both **VII** and **VIII** coexisting in equilibrium cannot be ruled out. The equilibrium constant in each special case would depend, of course, on the particular substituents attached to the functionalities involved.

Their cyclizations would then lead to the observed oxazolines of mixed stereochemistry. The relative proportion of **II** and **III** should reflect the equilibrium composition of **VII** and **VIII** because the ring closure of cations **IX**, and **X** to **II** and **III** is assumed to be faster than the C–C bond rotation that would furnish cis–trans isomerization.

A quite different result is obtained when the reaction conditions are modified. For example, when compound **XIII** is taken up in hot benzene containing a catalytic amount of *p*-toluenesulfonic acid, not only is an oxazoline obtained in which the absolute configuration of the unsymmetrical carbon (starred atom of Scheme 14.3) in the starting aziridine is retained (only 1% inversion), but the structurally isomeric oxazoline **XV** is produced also.[5]

This result calls for the intervention of a concerted mechanism, actually a 1,3-sigmatropic rearrangement where intermediates do not display significant charge separation.[6] The lack of stabilization by a relatively nonpolar solvent such as benzene probably plays a part in making the concerted rearrangement more favorable.

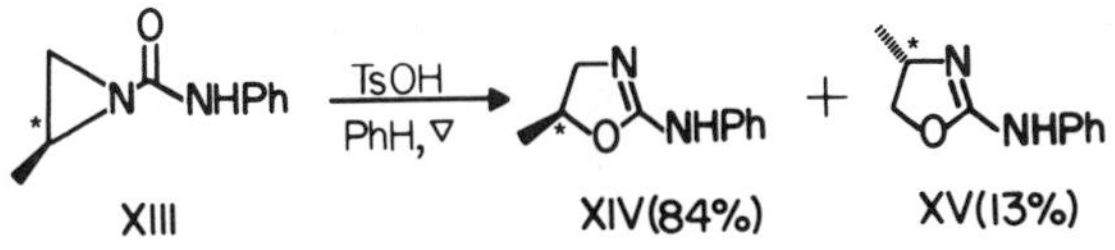

XIII —TsOH / PhH, ∇→ XIV(84%) + XV(13%)

SCHEME 14.3

REFERENCES

2. G. D. Andrews and J. E. Baldwin, *J. Am. Chem. Soc.*, **98,** 6705 (1976).
3. H. W. Heine, M. E. Fetter, and E. M. Nicholson, *J. Am. Chem. Soc.*, **81,** 2202 (1959).
4. C. U. Pittman, Jr. and S. P. McManus, *J. Org. Chem.*, **35,** 1187 (1970); G. A. Olah and P. J. Szilagyi, *J. Am. Chem. Soc.*, **91,** 2949 (1969).
5. T. Nishiguchi, H. Tochio, A. Nabeya, and Y. Iwacura, *J. Am. Chem. Soc.*, **91,** 5835 (1969).
6. For an interesting discussion on the transition from a stepwise to concerted mechanism in [2+4] cycloadditions, see A. B. Padias, S. T. Hedrick, and H. K. Hall, Jr., *J. Org. Chem.*, **48,** 3787 (1983).

PROBLEM 15

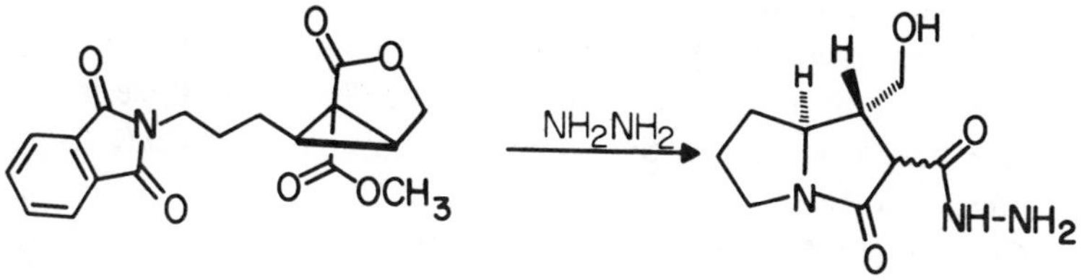

NH_2NH_2

I

II

1. S. Danishefsky, R. McKee, and R. K. Singh, *J. Am. Chem. Soc.*, **99**, 4783 (1977).

PROBLEM 15
Hydrazine Used for Ring Mutation

When chemical transformations involve the *quid pro quo* construction of cyclic structures at the expense of other cyclic portions of the molecule, the term ring mutation seems appropriate.[2] The present case is, visibly enough, an example of one such mutation where an azapentalenone derivative arises from an apparently unrelated γ-lactone fused to a cyclopropane.

The distracting paraphernalia that surrounds compound **I** may be considerably simplified by realizing that on one hand, the phthalimido and lactone rings disappear in the process, probably under the attack of the strongly nucleophilic hydrazine. This is suggested by the presence of hydroxymethylene and acylhydrazide groupings in **II.**

On the other hand, the count of C–C and C–N bonds in the backbone of starting and target materials indicates that two new carbon–nitrogen bonds are being formed, one of these necessarily at the sacrifice of the three-membered ring because there is no other active source of breakable C–C bonds. The other would come from the attack of nitrogen on methyl ester, and not the lactone. This may be inferred if one looks carefully at the cis configuration of the angular proton with respect to the hydroxymethylene substituent of **II.**

This prelude allows for a chain of events that would start with the release of the phthalimido group and the opening of the lactone ring. The liberated amine would become a powerful nucleophile five atoms removed from the nearest carbon of the cyclopropane fragment in intermediate **III** (see Scheme 15.1). This carbon atom is particularly electrophilic in character as the consequence of the 1,1-diacyl substitution.[2,3] In turn, this arrangement results in greater charge separation in the transition state derived from the cyclopropane ring opening, owing to the charge delocalization of the negative end,[4] although this interpretation has been challenged (see Problem 16).

As a consequence, the pyrrolidine derivative **IV** will be produced in as much as the nucleophilic attack of nitrogen is restricted to the carbon atom bearing the propylamino chain (spiro mode) since the analogous interaction with the other tertiary cyclopropyl carbon (fused mode, see dotted arrow) should afford the unrecorded piperidine derivative **V.** In actual fact, the spiro mode generally prevails over the latter.

Finally, the construction of the γ-lactam ring follows the standard interaction between the amino and ester groupings in **IV.**[5]

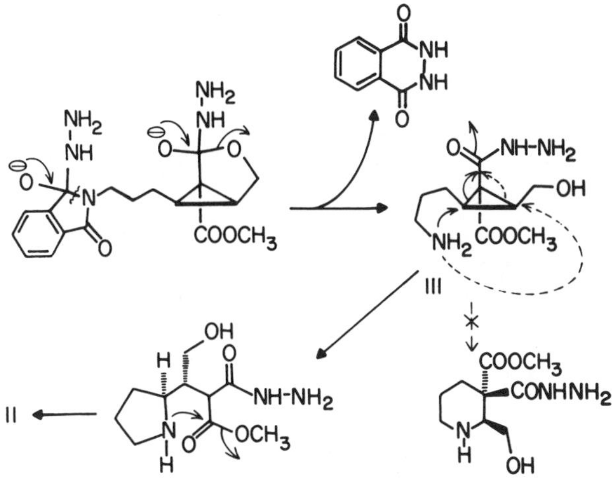

SCHEME 15.1

REFERENCES

2. This interesting circumstance was first recognized 90 years ago. See W. A. Bone and W. H. Perkin, *J. Chem. Soc.*, **67,** 108 (1895).
3. This subject has been reviewed. See S. Danishefsky, *Acc. Chem. Res.*, **12,** 66 (1979). See also T. Livinghouse and R. V. Stevens, *J. Am. Chem. Soc.*, **100,** 9479 (1978).
4. This rationale is complimented by my observation in connection with the much increased sensitivity of 1,1-diacylcyclopropanes towards nucleophiles by the introduction of electron-donating substituents on C-2, results that await publication.
5. For closely related references, see J. M. Stewart and H. H. Westberg, *J. Org. Chem.*, **30,** 1951 (1965); J. E. Dolfini, K. Menich, P. Corliss, R. Cavanaugh, S. Danishefsky, and S. Chakrabartty, *Tetrahedron Lett.*, 4421 (1966); E. J. Corey and P. Fuchs, *J. Am. Chem. Soc.*, **94,** 4014 (1972); G. Daviand and Ph. Miginiac, *Tetrahedron Lett.*, 997 (1972); E. W. Yankee, B. Spencer, N. E. Howe, and D. J. Cram, *J. Am. Chem. Soc.*, **95,** 4220 (1973); N. E. Howe, E. W. Yankee, and D. J. Cram, *J. Am. Chem. Soc.*, **95,** 4230 (1973). See also problem 17.

PROBLEM 16

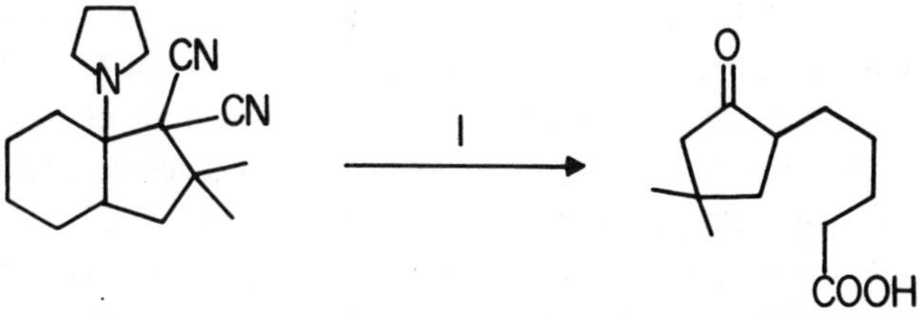

I

II(83%)

I = 85% H_2PO_4, HOAc, 150°, 71 hr.

1. W. F. Berkowitz and S. C. Grenetz, *J. Org. Chem.*, **41**, 10 (1976).

PROBLEM 16
A Drastic Ring Rupture and Recombination That Erases Most Molecular Clues

In view of our previous discussion on cyclopropane chemistry, it will be of some interest to glance briefly over the synthesis of compound **I.** Fragmentation of this molecule as shown in Scheme 16.1 illustrates the fact that **I** may be built by means of some sort of [3+2] cycloaddition. The two-carbon portion could be, for example, an enamine, and the three-carbon fragment a 1,3-dipolar structure.

This 1,3-dipolar compound does not reveal its dipolar nature until a second glance, for it is 2,2-dimethyl-1,1-dicyanocyclopropane. The double reactivity pattern that is illustrated by the simultaneous production of **I** and **VII,** is the result of either its tendency to undergo S_N2 attack with little charge separation in the transition state that leads to compound **I,** or to become zwitterionic intermediate **V** (that yields **VII**), which is effectively stabilized by the powerful electron withdrawing *gem*-dicyano unit (see Scheme 16.2). That the latter option does not contribute significantly to the reaction progress in nonpolar solvents is inferred from the exclusive formation of **I** in xylene (46% yield).

In contrast, a considerable body of evidence supports the existence of highly polarized intermediates in ring fragmentations of cyclopropane derivatives of type **IX.**[2]

The intelligent conclusion then is that these cyclopropanes are capable of ambivalent behavior, and the dielectric constant—or solvent strength or polarity—of the solvent employed has a strong bearing on the course of their reac-

N CN CN NC CN N NC CN

SCHEME 16.1

SCHEME 16.2

tions. These facts are to be kept in mind in the mechanistic considerations that explain this problem.

Contrary to previous exercises, none of the structural characteristics of the starting material can be identified in the product, save perhaps for the *gem*-dimethyl unit. Also, while **I** includes 9 *skeletal* carbons (this word is used to exclude those carbon atoms in the substituents in order to simplify the analysis of structural changes), target compound **II** contains 10. Obviously, this requires the incorporation of one functional carbon from the peripheral substituents, conceivably one of the nitriles, to the molecular backbone. This in turn implies that the five-membered ring of **II** is not the same cyclopentane unit of **I.**

The drastic changes that are implicit in this discussion are compatible with the rather harsh reaction conditions employed in this experiment. In fact, instead of limiting alternatives, strongly acidic media in highly polar solvents (acetic acid) facilitate the numerous pathways open to thermal induction and

IX

R^1, R^2= ALKYL, VINYL, ARYL, H

R^3, R^4= CN, COOR, SO_2R

proton catalysis. As a consequence, the postulation of more than one mechanistic sequence in this sort of situation is not only common but necessary.

Compound **I** must be activated by the strongly acidic medium. One possible protonation site is the basic pyrrolidinium nitrogen atom (structure **X**), which would favor the formation of carbenium ion **XI** (see Scheme 16.3). This intermediate would gain additional stabilization from the polar solvent. Conditions for an S_N1 substitution would prevail to yield **XII,** whose fragmentation would resemble, formally speaking, the retro cycloaddition process used in the synthesis of **I** (i.e., **VIII** → **I**). The resulting ketone (**XIII**) would contain the elements required for the construction of a five-membered carbocycle with the desired incorporation of the functional carbon of one of the nitrile groups, as

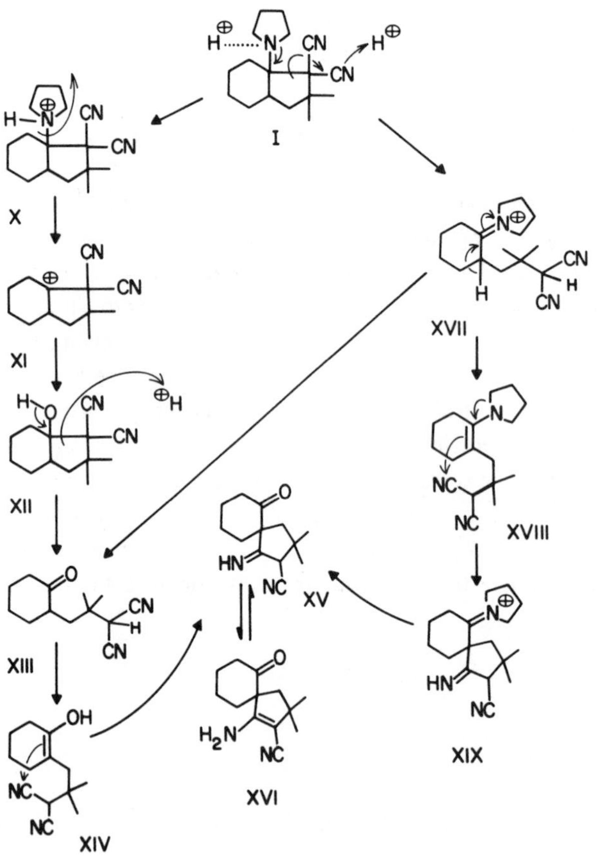

SCHEME 16.3

shown in **XIV.** The resulting keto imine (**XV**) presumably would coexist in equilibrium with its enamino form **XVI.**

Alternatively, a thermal activation of the starting material **I** would follow the reverse course of its synthesis to furnish cation **XVII,** which would then undergo a similar spiro cyclization by way of enamine **XVIII.** The hydrolysis of the ensuing iminium ion would lead directly to the key intermediate **XV.**

The pathway from the equilibrium mixture **XV–XVI** to the target compound **II** may well split again to follow a ketonic-type cleavage either via nitrile ketone hydrate (**XX** → **XXI**), or by a shorter route that requires the intervention of **XXVII** (see Scheme 16.4).

Such mechanisms are far too intricate to allow for a formal mechanistic

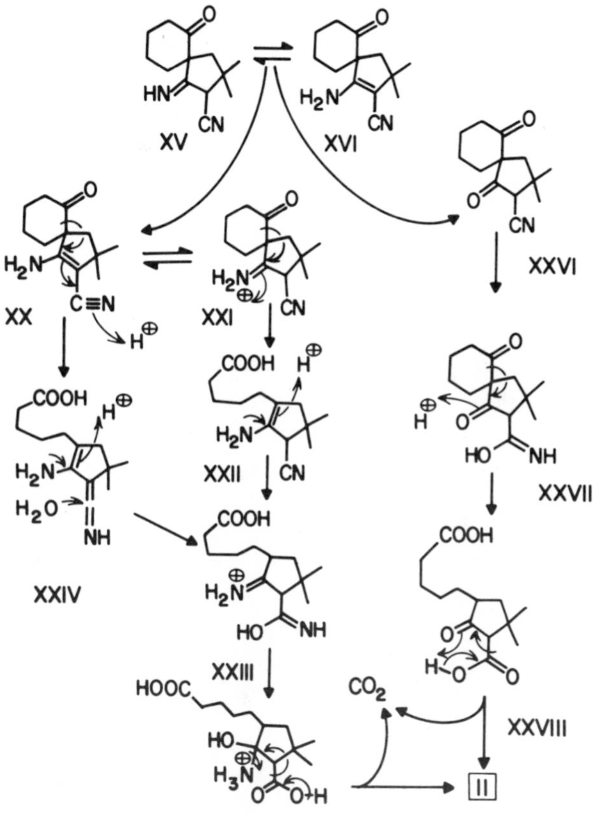

SCHEME 16.4

study, and a responsible choice is thus difficult. However, the sequence shared by intermediates **X** and **XIII** involves the solvolysis of an amine, which is a relatively rare occurrence for tertiary amines. The alternative route represented by **XVII** is probably a better choice to arrive at key intermediate **XV,** since it follows well established Mannich chemistry. As for the progress of **XV** towards product **II** (Scheme 16.4), there is no strong argument against the participation of any of the postulated intermediates.

REFERENCES

2. See, for example, A. B. Chmurny and D. J. Cram, *J. Am. Chem. Soc.*, **95,** 4237 (1973).

PROBLEM 17

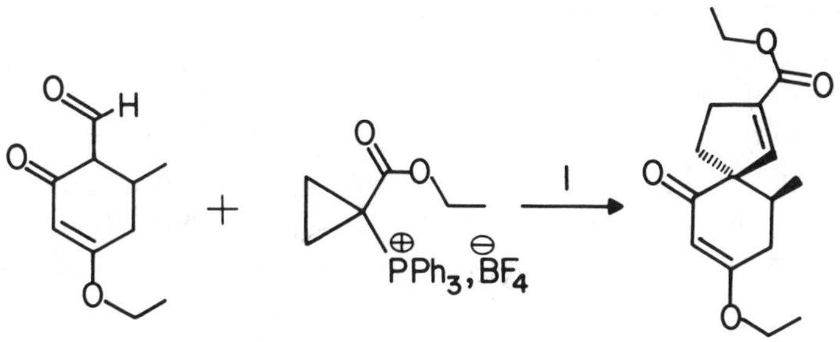

I

II

III

I NaH, HMPA

1. W. G. Dauben and D. J. Hart, *J. Am. Chem. Soc.*, **97**, 1622 (1975).

PROBLEM 17
In Situ Creation of a Phosphorous Ylide and Construction of Disparate Bonds

In this problem, the count of atoms in starting and target materials indicates that compounds **I** and **II** must somehow become coupled, and this coupling is accompanied by loss of the triphenylphosphonium moiety of **II** and the oxygen atom of the aldehyde function.

Fragmentation of the target compound in terms of the starting materials can be done only by separation of the five-membered ring as indicated by dotted lines in Scheme 17.1. While it is easy to perceive where fragment **IV** comes from, the origin of the butyric ester fragment requires further comment.

Fragment **VI** is coupled to **IV** at two points. In the first of these two, the α carbon of the ester must be such that a C=C bond is built. This task could be accomplished by, for example, a phosphorous ylide or phosphorane that would react with the aldehyde portion of **I** in a manner reminiscent of the Wittig reaction. This implies the existence of a structure such as **VII** at some point of the reaction scheme.

The second coupling point of the butyric ester fragment **VI** requires that the γ carbon of this ester is so functionalized as to allow for a C–C bond to be formed. How is this performed?

From our previous discussion about electrophilic cyclopropanes (see Problem 15), it should be remembered that a marked reduction of electron density in the three-membered ring occurs when strong electron withdrawing groups are substituted either directly or in conjugation with the trimethylene ring; and

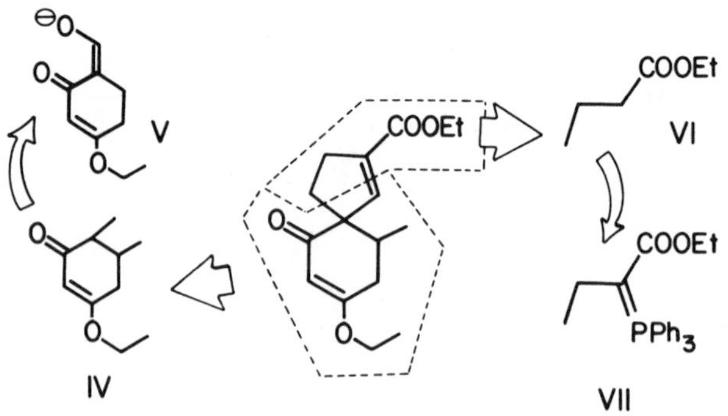

SCHEME 17.1

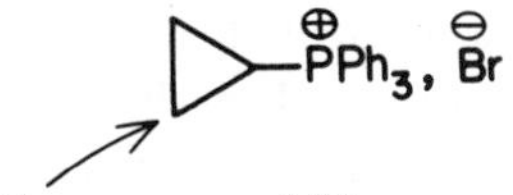

VIII

there is the triphenylphosphonium function. In fact, that this group imbues cyclopropane rings with a moderate electrophilic character was first demonstrated by the reaction of the bromide salt **VIII** with the sodium salt of salicylaldehyde,[2] although yields were not satisfactory for synthetic purposes.

The electrophilicity of **VIII,** nevertheless, was dramatically enhanced by the introduction of a second electron withdrawing substituent on the same carbon that bears the triphenylphosphonium unit. Professor Fuchs solved this synthetic problem by showing that cyclopentenes could be obtained in high yields from **II,**[3] a compound that has become commercially available as a result.

If compound **II** is electrophilic, the involvement of the γ carbon of **VI** now becomes clear. Thus, the sodium hydride generated anion **V** may be imagined as attacking nucleophilically carbon C-2 of activated cyclopropane **II.** The resulting anion is precisely the proposed phosphorous ylide postulated as structure **VII** by our previous fragmentation analysis. What follows then is an intramolecular Wittig reaction with the departure of triphenylphosphine oxide that was predicted by the atom budget procedure (see Scheme 17.2).

It is noteworthy that none of the epimeric spiroketone **IIIb** was detected.

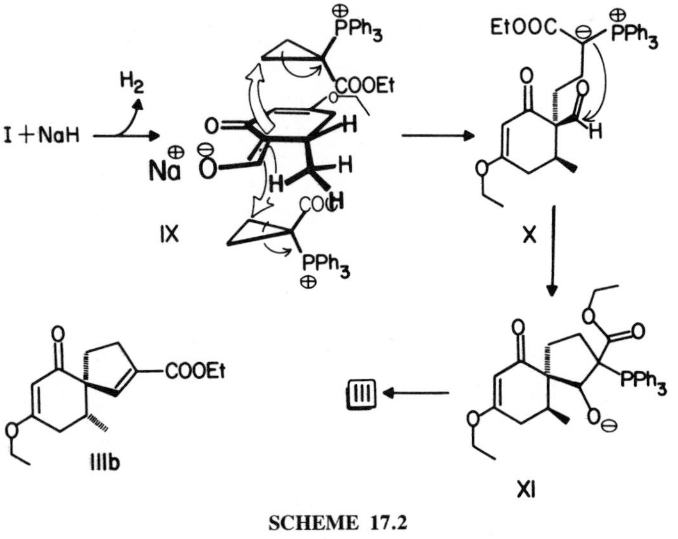

SCHEME 17.2

"The stereoselectivity of the attack of the enolate on the activated cyclopropane most likely is due to steric factors arising from the preferred pseudoaxial conformation of the incipient methyl group[1]". In other words, the approach of the rather bulky phosphonium salt **II** is less perturbed by the methyl group when it rests on the opposite side of the molecule in the transition state, presumably in the pseudoaxial conformation, as portrayed in **IX.** This remarkable stereoselectivity has been put to good use in the synthesis of some natural products of the spirovetivane family.[4-6]

REFERENCES

2. E. E. Schweizer, C. J. Berninger, and J. G. Thompson, *J. Org. Chem.*, **33,** 336 (1968).
3. P. L. Fuchs, *J. Am. Chem. Soc.*, **96,** 1607 (1974).
4. See, for example, W. G. Dauben and D. J. Hart, *J. Org. Chem.*, **42,** 922 (1977).
5. For a review on synthetically useful homologs of **II,** namely, Cyclopropyl sulfonium ylides, see B. M. Trost, *Acc. Chem. Res.*, **7,** 85 (1974).
6. For rewarding and informative reading on spiro cyclization see P. A. Krapcho, *Synthesis*, 383 (1974).

PROBLEM 18

I + II $\xrightarrow[\Delta,\ 20hr]{NaOCH_3 \quad HOCH_3}$ III

I: 1-[(dimethylamino)methyl]-2-naphthol, with $N(CH_3)_2$ and OH labels
II: PhCO–CH2–SO2Ph (O, O, S, O labels)
III: 1-(2-phenylsulfonylethyl)-2-naphthol, with OH and SO_2Ph labels

1. K. K. Balasubramanian and S. Selvaraj, *Synthesis*, 138 (1980).

PROBLEM 18
One-Carbon Elongation of Side Chain Aided by Fragmentation of the Main Ring

This self-proclaimed "simple elegant" method for the preparation of 2-hydroxy-substituted aromatic phenyl ethyl sulfones, a family of compounds with insecticidal activity, is actually the unexpected result of an attempted synthesis of sulfonyl chromenes of type **IV** (see Scheme 18.1). These dihydropyranyl derivatives have been prepared by the reaction of α-hydroxybenzylamines with active methylene compounds such as malonic, cyanoacetic, and acetoacetic esters as well as 1,3-diketones, under alkaline conditions.[2]

The transformation of the naphtholic Mannich base **I** into **III,** formally a one carbon homologation of the alkyl side chain, represents a deviation from the general behavior of the chromene synthesis. Mechanistically, it is closely related to the normal process, however. For one thing the active methylene component **II,** which forms a very stable carbanion owing to the combined electron withdrawing effects of the phenyl sulfone and benzoyl units, becomes coupled with **I** by way of an unorthodox displacement of the dimethylamino group, which is classically deemed as a poor leaving group in its uncharged state. The departure of this fragment may nevertheless be aided by the favorable proton abstraction from the acidic O–H on naphthol with the consequential increase in electron density of the aromatic ring. Here a six-membered ring to account for the δ-elimination process is conceivable. Assistance by the protic solvent is also to be expected. In addition, cross-conjugated ketone **V** could well result from this elimination prior to the nucleophilic attack of the anion derived from **II.**

Intermediate **VI,** produced either by direct nucleophilic displacement or 1,4 addition depending on whether **V** is the actual substrate, may then be transformed into the final product. This may occur either by way of an acidic-type fragmentation via the bimolecular attack of the methoxide anion on the benzoyl

OH NR$^3_2$ + O R^1 ⊖ R^2 $\xrightarrow{B^{\ominus}}$ O R^1 R^2 IV

SCHEME 18.1

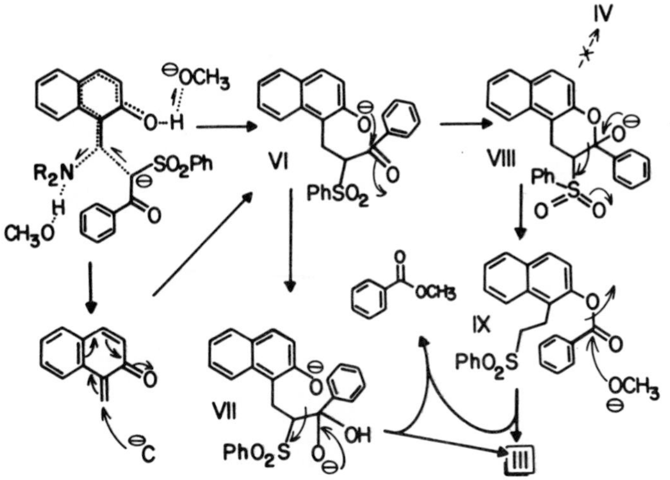

SCHEME 18.2

group, namely, **VI** → **VII,** or conversely by means of an analogous intramolecular cyclization, but with intervention of the naphthoxide oxygen atom, a sequence that would yield the chromene precursor (**VIII**) (see Scheme 18.2).

In our view, the latter route is probably preferred because the proximity of the reacting groups should be less energy demanding than the effective collision of the methoxide anion required by the bimolecular route. Also, besides being aesthetically appealing, earlier experience[2] on the synthesis of chromenes from 1,3-dicarbonyl compounds provides a strong indication of a cyclization step prior to the final fragmentation. The benzoyl ester **IX** emerges from this second pathway, which meets the end of the sequence by a now necessary bimolecular benzylic-type fragmentation.

REFERENCES

2. H. Hellmann and J. L. W. Pohlmann, *Justus Liebigs Ann. Chem.*, **643,** 38 (1961) and references cited therein. See also K. Friederich and H. Kreuscher, *Angew. Chem.*, **72,** 780 (1960).

PROBLEM 19

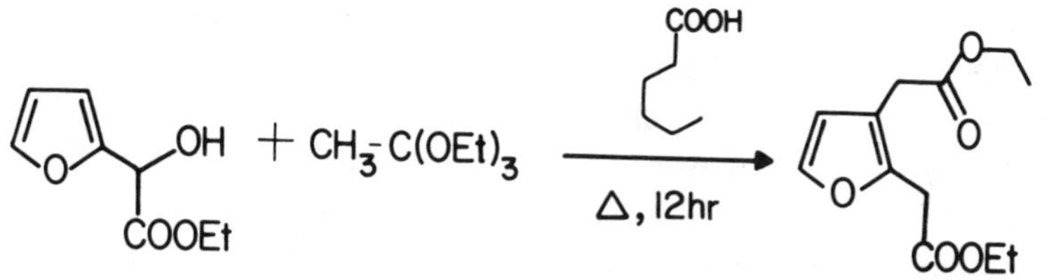

OH + CH_3-$C(OEt)_3$ → (COOH, propionic acid; Δ, 12hr)

COOEt

I

COOEt

II(75%)

1. S. Raucher, A. S.-T. Lui, and J. A. Macdonald, *J. Org. Chem.*, **44**, 1885 (1979).

PROBLEM 19
Claisen-Type Rearrangement of a Mixed Ketal Intermediate

The synthesis of the target compound by this reaction involves the introduction of an acetic ester residue at C-3 without any apparent activation of the furan nucleus other than the somehow invigorating effect of the C–OH bond rupture, a group that is in fact absent in **II.** That is, the oxidation level of the carbinol carbon must be used up in some way to make the new C–C bond at C-3.

One rather rudimentary way of applying this oxidation level transfer would be to pull off the hydroxide unit from **I** with the aid of acid and attack the resulting carbenium ion in a S_N1' fashion with diethyl keteneacetal, which in turn would be derived from the acid catalyzed elimination of ethanol from the starting triethyl ortho acetate. Alternatively, a S_N2' process, in very much the same terms, would also provide an unsophisticated route to **II** (see Scheme 19.1). Let us add now some further experimental evidence and see whether the proposed mechanism is still valid.

Although no one has ever been able to trap any intermediate in the several related examples discovered so far, some experimental facts of crucial importance to this mechanism have been produced, the most outstanding of which are

1. The reaction of the analogous aliphatic allyl ethers is stereoselective, an observation first applied[2] to the synthesis of trans-trisubstituted olefins, in particular squalene (see Scheme 19.2).

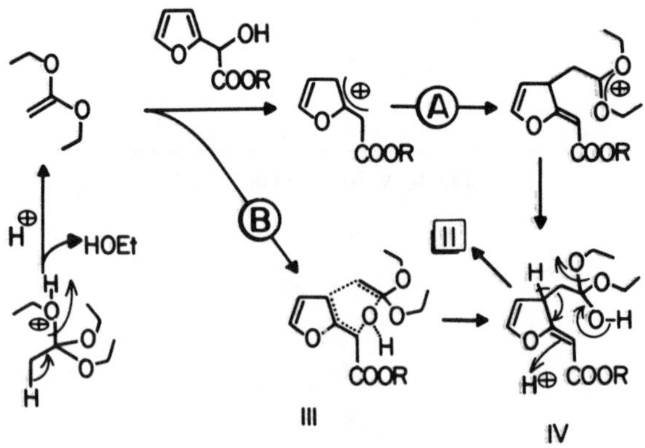

SCHEME 19.1

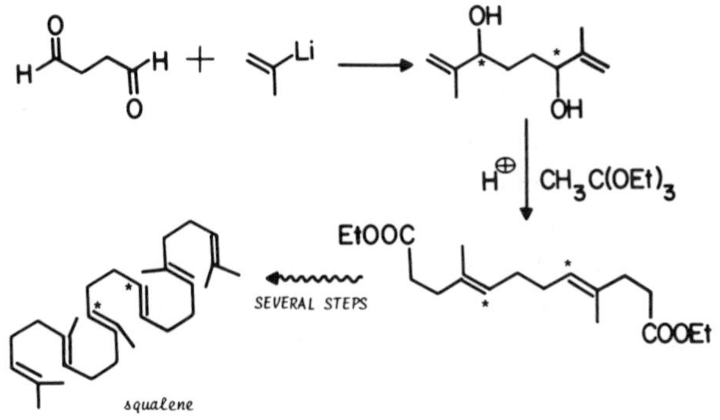

SCHEME 19.2

Amazingly, all central C=C bonds were 98% trans, a remarkable stereoselectivity for any reaction. This fact rules out the intervention of an S_N1' route—unless some inexplicable stereocontrol is in operation. Since the transition state must embody rotational restrictions, the S_N2' mechanism would not be invalidated. In this bimolecular approach the incoming and departing groups would be somehow associated as to discourage rotation of the carbon skeleton. Structure **III,** for example, would contain a comfortable six-membered transition state that should fulfill this requirement thoroughly, although it is unprecedented.

2. Tertiary alcohols such as **V** do not undergo this reaction according to some experimental evidence.[3] Any increase in the alkyl substitution at the carbinol carbon, however, would be expected to facilitate both S_N1' and S_N2' processes. Evidently, the basis of the nucleophilic displacement mechanism is not as solid as we originally thought.
3. *p*-Methoxybenzyl alcohol does not react at all.[4] Consequently, the nucleophilic attack becomes untenable.

There is a more elegant alternative that satisfies the rotational restrictions that ensure stereoselectivity. This alternative makes use of the

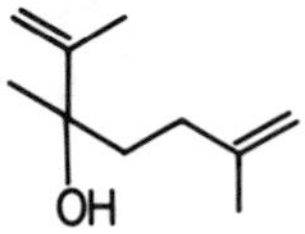

V

well-known Claisen rearrangement.[5] As we know from Problem 12, alcohols are easily coupled to ortho esters under acid catalysis[6] to furnish mixed ortho esters such as **VI.** This compound would then undergo acid assisted β elimination of ethanol to yield the ketene acetal **VII.** If we imagine for a moment that the furan nucleus of **I** is a benzene ring and compare the intermediate **VII** with the aromatic esters used for the classical version of the Claisen rearrangement, it will become apparent that both situations are quite similar, save that the former is a vinyl benzyl ether and the Claisen rearrangement substrate is a phenyl allyl ether instead (see Scheme 19.3). This is indeed a difference of some importance since the latter smoothly undergoes rearrangement while the vinyl–benzyl combination does not.

The apparently innocuous ethoxycarbonyl group in **I** must have a pivotal role in facilitating the rearrangement since other vinyl benzyl ethers with this group readily undergo the electrocyclic reaction to introduce the acetic ester residue at C-3 of compound **I.**[4] Such an effect of the carbethoxy group has been attributed "to the increased stabilization of the putative intermediate **IX** formed by the [3,3] sigmatropic rearrangement of [intermediate **VII**],[1]" a contention that requires a late transition state. This interpretation has been termed the Claisen ortho ester rearrangement, and is presently referred to by that name.[7,8]

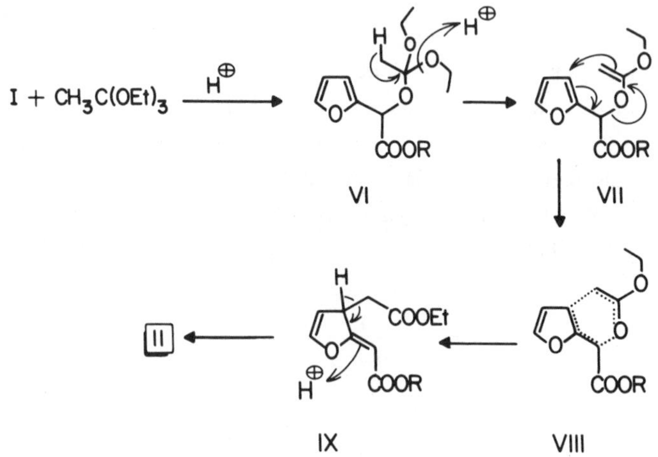

I + $CH_3C(OEt)_3$ $\xrightarrow{H^{\oplus}}$ VI → VII → VIII → IX → II

SCHEME 19.3

REFERENCES

2. W. S. Johnson, L. Werthemann, W. R. Bartlett, T. J. Brocksome, T. Li, D. J. Faulkner, and M. R. Petersen, *J. Am. Chem. Soc.*, **92,** 741 (1970).
3. K. A. Parker and R. W. Kosley, Jr., *Tetrahedron Lett.*, 691 (1975).
4. S. Raucher and A. S.-T. Lui, *J. Am. Chem. Soc.*, **100,** 492 (1978).
5. For a review see S. J. Rhoads and N. R. Raulins, "The Claisen and Cope Rearrangements," *Org. React.*, **22,** 1 (1975).
6. R. H. DeWolfe, *Carboxylic Orthoacid Derivatives*, Academic, New York, 1970.
7. It is a good didactic exercise to follow the development of the topic treated in Problem 12 as an offshoot of the present subject. Read in order: (a) References about the [3,3] sigmatropic rearrangement given in reference (4) of Problem 12; (b) reference (4) of Problem 12; (c) K. A. Parker and R. W. Kosley, Jr., *Tetrahedron Lett.*, 3039 (1977) and 341 (1978).
8. See also R. G. Salomon, *Tetrahedron Lett.*, 3235 (1977) and 3199 (1978).

PROBLEM 20

1. M. C. Vander Zwan, F. W. Hartner, R. A. Reamer, and R. Tull, *J. Org. Chem.*, **43**, 509 (1978).

PROBLEM 20
A Stepwise Electron Transfer That Requires Two Moles of Oxidant

Structures **I** and **III** are members of a vast class of β-lactam antibiotics around which much research is being conducted in many major industrial laboratories. Their chemistry is far from trivial since their multiple functionality demands careful handling.

In the present reaction the impressive complexity of compound **I** boils down to a much simpler system by concentrating our attention only on that sector of the molecule that undergoes change, that is, the terminal α-amino acid. Although the remaining reactive centers should be kept in mind, these represent a distraction and therefore will be disregarded.

This reaction is an unexpected deviation from an attempted oxidation of the α-amino acid **I** to the corresponding aldehyde (**V**) (see Scheme 20.1). This expected reaction would have followed the guidelines of the Strecker degradation,[2] whereby 1 mol of quinone is normally used.

By contrast, 2 mol of quinone were consumed in the production of the benzoxazole derivative **III** (see Scheme 20.2). This is evidenced by the incorporation of a molecule of quinone **II** in **III** (in the benzoxazole unit) and the recovery of one additional mole of **IV.** In this process the α carbon of the carboxylic acid chain is oxidized from the stage of an amine to that of an imino ester. Compare Schemes 20.1 and 20.2.

In essence valence electrons must be going from **I** to **II** in two successive stages. It is instructive to analyze these electron transfers in some detail.

The interaction of **I** and **II** may be examined within the framework of imine generation from amines and carbonyl compounds. The alkaline environment should secure the nucleophilicity of a neutral amino function by deprotonation

SCHEME 20.1

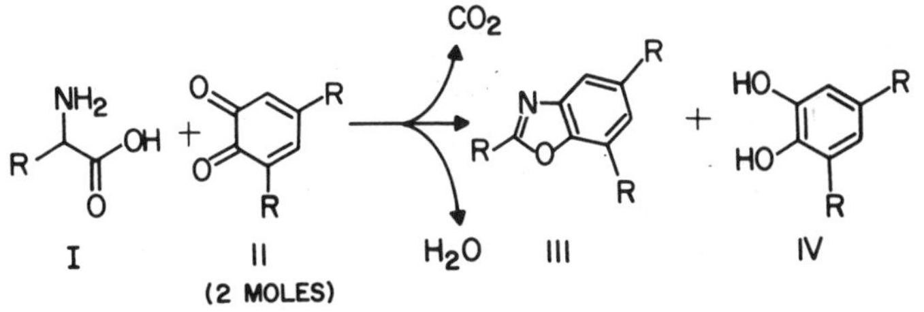

SCHEME 20.2

of the carboxylic acid and by the elimination of water in **VII** to yield the imine derivative **VIII** (see Scheme 20.3). If the C=N bond in this intermediate is considered a C=C bond for a moment one will realize that this is an opportune stage for the liberation of carbon dioxide, because this would be a vinylogous version of the well-known decarboxylation of β-keto acids. During this step the carbon chain of **I** is oxidized to the level of an aldimine but at the expense of the loss of a C–C bond. That is, the oxidized species at this stage is not the global aminoquinone backbone **IX,** which in actual fact is reduced. The oxidized species is carbon dioxide instead. In other words, valence electrons are flowing from carbon dioxide to the molecular backbone.

Aldimines such as **IX** are labile towards nucleophiles much in the way aldehydes are, and the nonoxidative intramolecular coupling with the phenoxide oxygen would be expected to follow. The gap between the resulting interme-

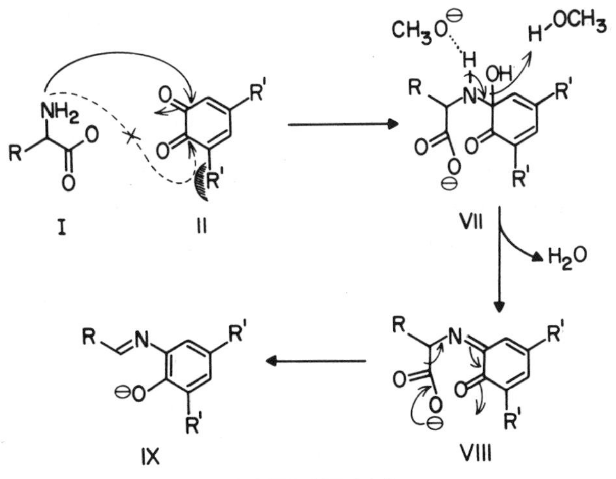

SCHEME 20.3

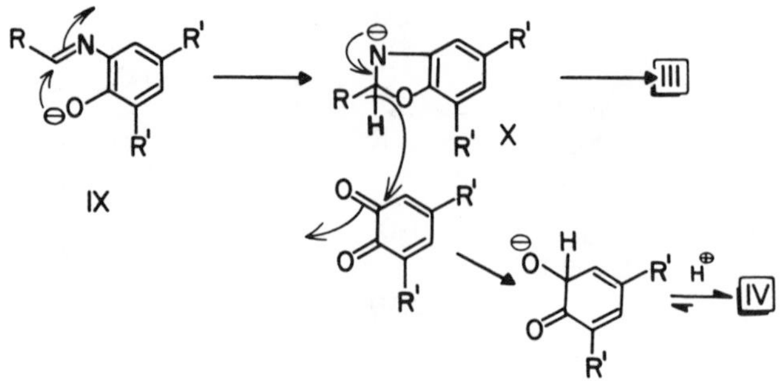

SCHEME 20.4

diate **X** and the target compound entails only the formation of a C=N bond (see Scheme 20.4). However, **X** contains too many electrons to accomplish this without an external electron sink (and oxidizer). Excess electrons are ejected from **X** in the form of a hydride anion that is conveniently captured by the second mole of quinone **II.** Alternatively, successive one electron transfers with intervening proton delivery would also be conceivable since quinones are also effective substrates for one electron oxidations.[3] This would become the second oxidative step.

In short, the aminobutyric chain of **I** is oxidized to the level of an ester and the terminal carboxylic acid carbon to the level of carbon dioxide. This implies a grand total of three oxidative operations (not two), two on account of 2 mol of quinone, and one because of the C–COOH bond disconnection.

REFERENCES

2. A. Strecker, *Justus Liebigs Ann. Chem.*, **123,** 363 (1862); A. Schoenberg, R. Moubasher, and A. Mostafa, *J. Chem. Soc.*, 176 (1948).
3. For recent reviews see L. Eberson, *Adv. Phys. Org. Chem.*, **18,** 79 (1982); A. Pross, *Acc. Chem. Res.*, **18,** 212 (1985).

PROBLEM 21

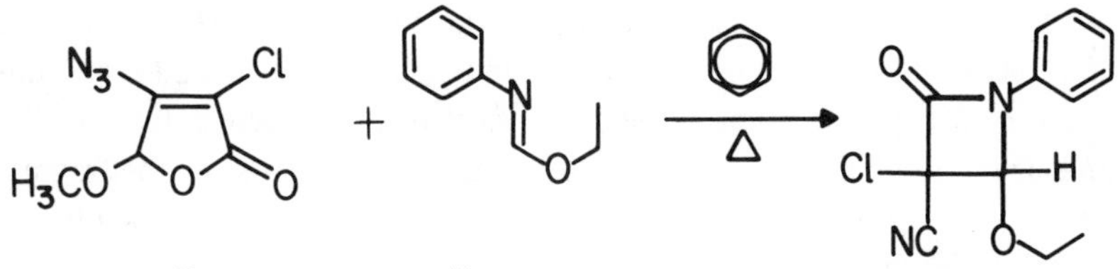

I II III

1. H. W. Moore, L. Hernandez, and A. Sing, *J. Am. Chem. Soc.*, **98,** 3728 (1976).

PROBLEM 21
A β-Lactam Synthesis Via a Deep Rooted Dislocation of the Starting Material

Now that our attention has been focused on the β-lactam structure of the preceding problem, a rather unusual synthesis of this interesting ring constitutes the next problem. The fact that two carbon and two oxygen atoms are lost in the conversion of **I** into **III** suggests the occurrence of a deep rooted fragmentation of the starting compound. That the missing atoms are most likely those corresponding to the *acetal* sector of compound **I** is emphasized by the fragmentation scheme shown in Scheme 21.1. Here one sees that all the elements of *N*-phenyl-formimidate (**II**) are incorporated into **III**, hence only the carbonyl and the carbon atom bearing chlorine are remnants of **I** in the target product.

In addition, the carbon atom of the nitrile function of **III** must proceed from the starting lactone in order to maintain the atomic balance. This premise proves to be the starting point of the entire mechanistic sequence if one realizes that the azido group can accommodate those electrons released by the fragmentation suggested previously. In that case, molecular nitrogen would be liberated as shown in **IV**, the consequence of which would be zwitterion **V** (see Scheme

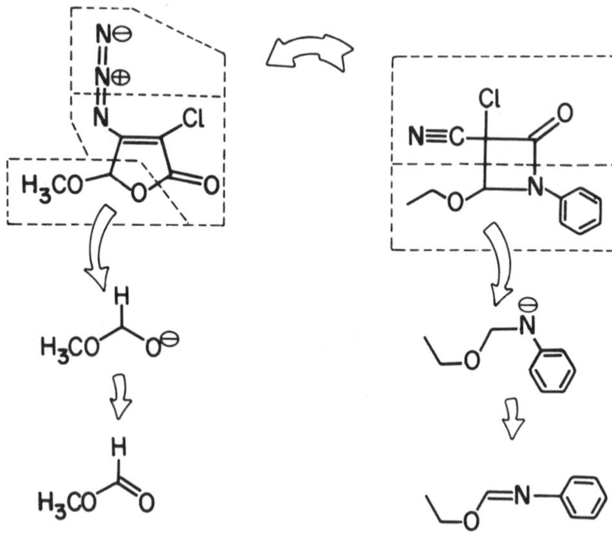

SCHEME 21.1

21.2). Although in principle this structure would be ideal for [2+4] cycloaddition with the polarized C=N bond of **II** to give a relatively stable six-membered azalactone **VII,** it would face some difficulties during its transformation to end product **III.** For one thing, if a concerted mechanism would operate, a rather forced boat conformation such as **VIII** (although not completely impossible) would be required to bring nitrogen and carbonyl groupings at interaction distance.

If, conversely, compound **VII** were to be treated as a cyclic amino-ortho ester, a nonconcerted, thermally induced C–N bond rupture that would lead to another zwitterionic structure (**IX**) and from there to **III** by way of recombination–fragmentation would also be more feasible, because ortho ester derivatives are prone to such thermal fragmentation (see Scheme 21.3).[2]

Still a third mechanism is conceivable if the key intermediate **V** is allowed

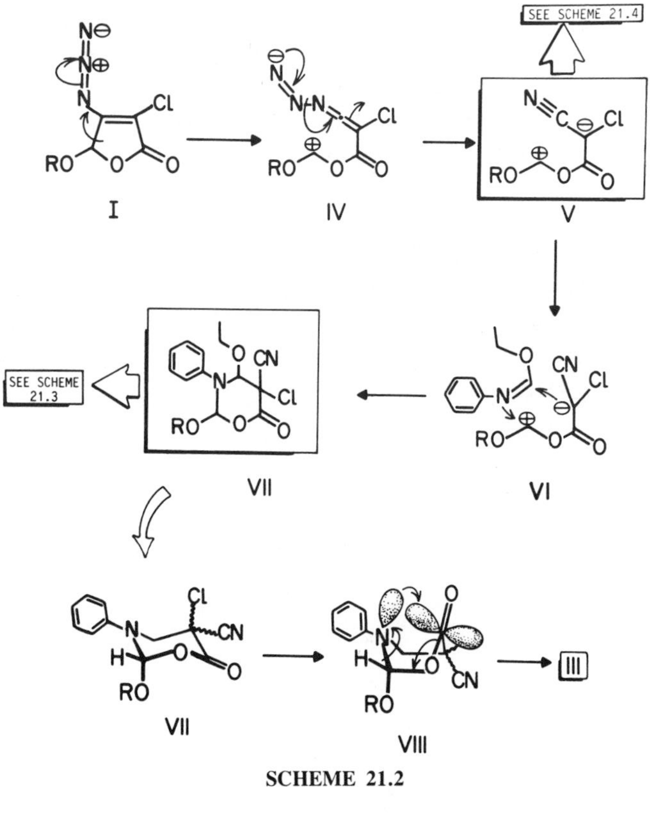

SCHEME 21.2

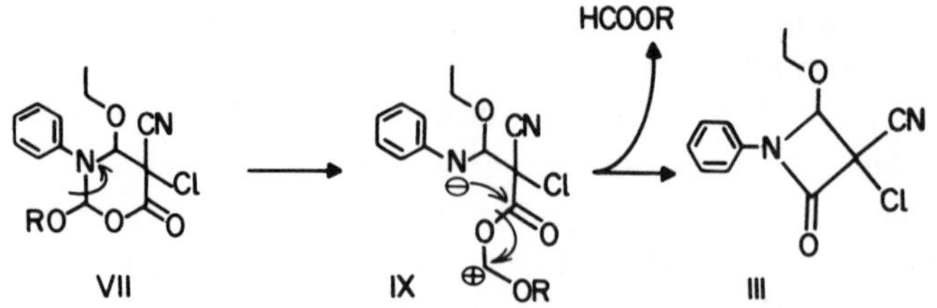

SCHEME 21.3

to undergo fragmentation in the manner indicated on Scheme 21.4. The outcome of this process is cyanochloroketene (**X**). This intermediate would then be capable of adding to **II** in a [2+2] fashion, very probably in a nonconcerted manner, via **XI.**[3]

Of these various propositions the one portrayed in Scheme 21.2 is the least attractive because of the geometric limitations imposed in the conversion of **VIII** to **III.** A definitive choice among the other proposals is difficult however, because these alternatives make due use of stabilizing functions such as nitrile and ester next to carbanions **V** and **XI,** and of other chemically sound processes.

Although this and other analogous reactions have not been studied mechanistically, some experimental evidence of relevance to mechanisms has been collected. This is summarized as follows:

(a) The thermolysis of **XII** gives imide **XIV** in 26% yield (see Scheme 21.5). While this is consistent with the intervention of a **V**-type inter-

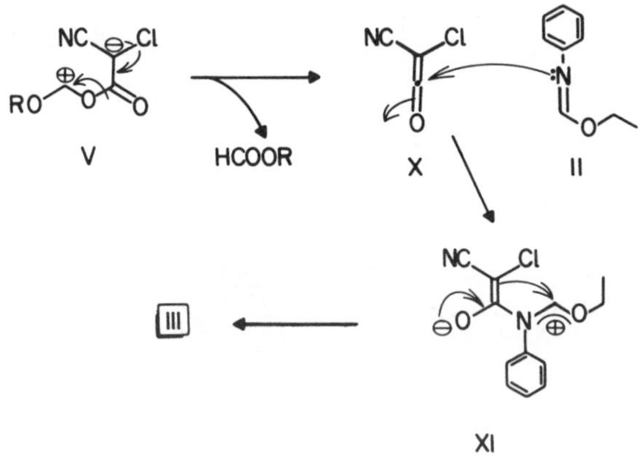

SCHEME 21.4

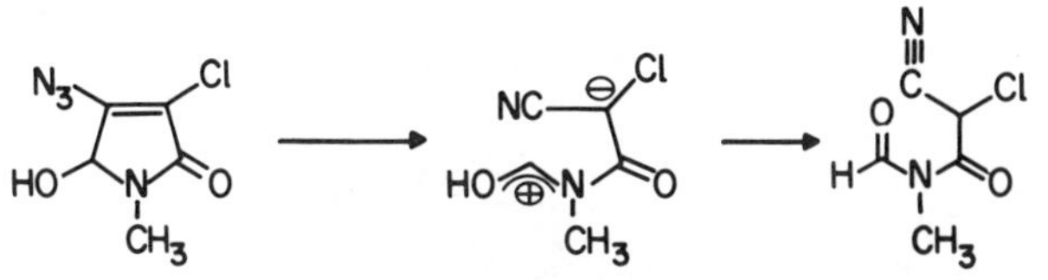

XII XIII XIV

SCHEME 21.5

mediate, such as **XIII,** and by now there is little doubt about its existence,[4] this datum alone adds nothing to disprove or invalidate the participation of ketene or aminolactone species.

(b) The thermolyses of **XV** and **XVI** have also been studied, and the results are illustrated in Scheme 21.6. This data leads to the same conclusion as in **(a)**.

(c) More informative is the thermolysis of **XVII** from which the β-lactam **XIX** was isolated as a single stereoisomer. When this experiment was repeated in the presence of dicyclohexylcarbodiimide (DCC) (**XX**), a very reactive compound that can much more effectively assume the role played by **II** in the capture of dipolar species, no β-lactam was ever detected. Adduct **XXI** was isolated instead.

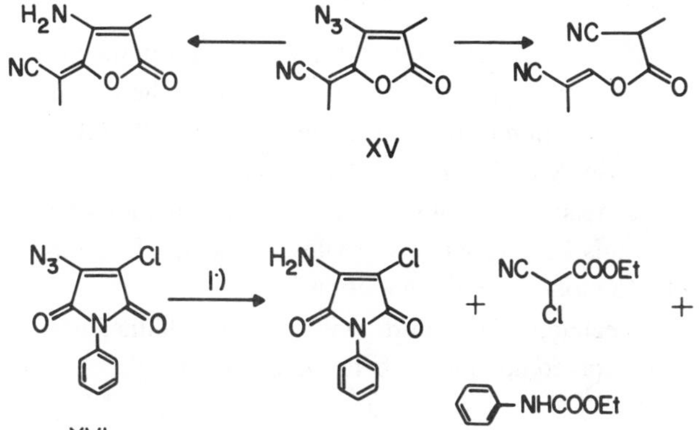

XV

XVI

I) = C_6H_4Cl, EtOH, Δ

SCHEME 21.6

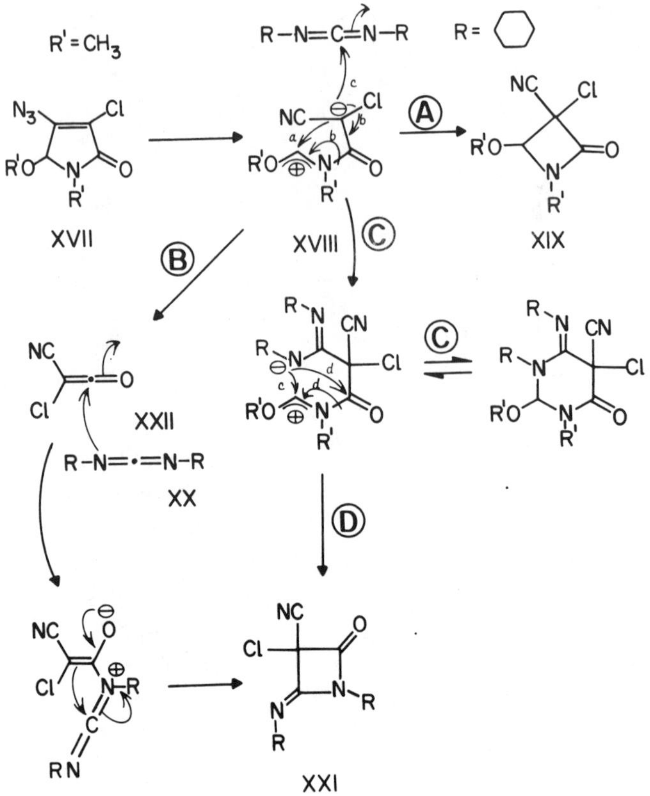

SCHEME 21.7

Along the same lines, thermolysis of the β-lactam in the presence of DCC gave the same adduct **XXI.** Although it seems that these experiments point towards the decomposition of the zwitterionic structure **XVIII** to ketene **XXII,** particularly because **XVIII** emulates **XI** perfectly, there are no elements of judgment in these reactions against the alternative hypothesis that bypasses the intermediacy of this ketene, namely, routes **C** and **D** of Scheme 21.7, which mimic the other options open to zwitterion **V.**

The actual mechanism, therefore, will remain undefined until the chemical behavior of **VII**-type compounds is better understood.

REFERENCES

2. See structure **VII** in Problem 12, Scheme 12.1.
3. In general, [2+2] cycloadditions are rarely concerted under thermal conditions, with

the exception of ketenes. Some degree of concert, however, can be achieved in the analogous photochemical transformation. The observed stereoselectivity in the formation of **III** can be accounted for by a nonconcerted, stereoelectronically controlled cycloaddition. See reference 1.

4. For the conceptually similar chemistry and thermolysis of azidoquinones, see H. W. Moore, *Chem. Soc. Rev.*, **2,** 415 (1973).

PROBLEM 22

F_3CSO_3H

CH_2Cl_2

$-70°$

I + II → III + IV

1. G. Buchi and C.-P. Mak, *J. Am. Chem. Soc.*, **99,** 8073 (1977).

PROBLEM 22
Divergent Cycloadditions from the Same Putative Intermediate

Fragmentation analysis of compounds **III** and **IV**—indicated in Scheme 22.1—reveals that unidirectional, regio- and stereospecific [3+2] and [4+2] cycloaddition processes, respectively, of the naturally occurring *trans*-isosafrole (**II**) are basically the mechanistic concepts underlying this transformation.

In mechanistic analysis, however, more important than what happens is how it happens. In this particular example, the second question is not devoid of difficulties.

If, on one hand, isosafrole is considered a moderately nucleophilic alkene, owing to the injection of electron density into the styrene structure by the oxygen substituent, the three electrophilic centers of compound **I** (denoted by asterisks) ostensibly do not participate as bonding centers in this reaction. The electrophilic or receptor role is somehow assumed in turn by the potentially

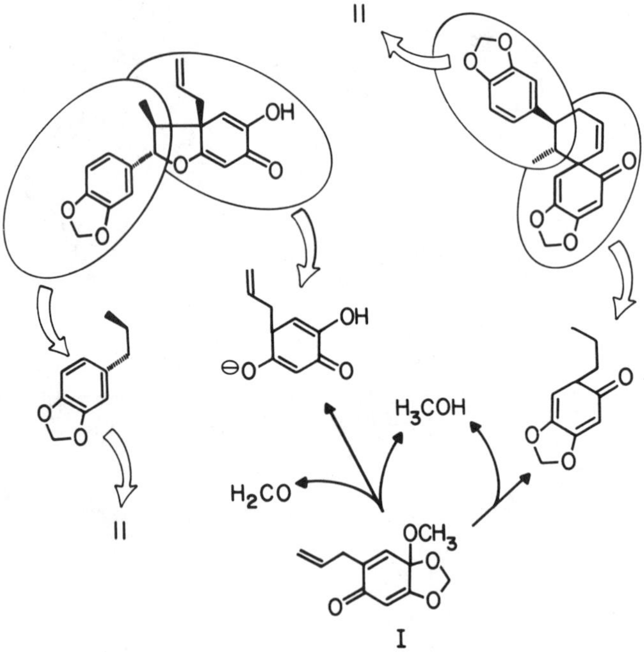

SCHEME 22.1

nucleophilic tertiary carbon α to the ketone and the terminal methylene of the side chain of substrate molecule **I.** Therefore, proper activation of these two carbons ought to be achieved by way of polarity inversion, particularly of the α carbon. This can be obtained by triflic acid assisted elimination of methanol from **I,** which leaves allylic carbenium ion **V.** This situation warrants the formation of the first C–C bond between the α carbon of *trans*-isosafrole (**II**) and the α carbon of **I.** This coupling would then lead to the relatively stable benzylic carbenium ion **VI,** which would be theoretically prone to evolve in two diverging ways, **A** and **B**, depending on whether it is trapped by the carbonyl oxygen or the terminal olefin. At the end of both pathways would be target compounds **III** and **IV** as portrayed in Scheme 22.2.

It can also be argued that the production of compound **IV** takes place by way of a concerted Diels–Alder type cycloaddition of **II** on diene **IX,** which in turn would be derived from 1,4 elimination of methanol from **I** (see Scheme 22.3). This process would comfortably accommodate the observed stereospecific-

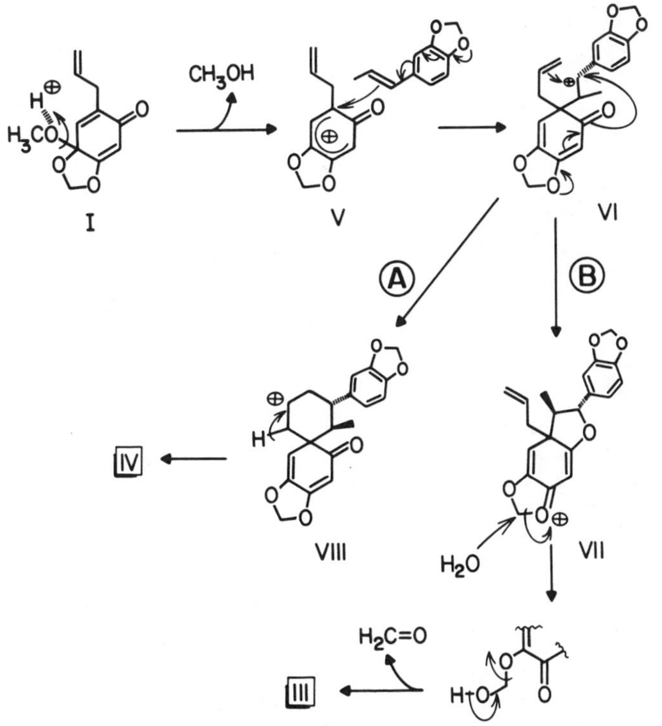

SCHEME 22.2

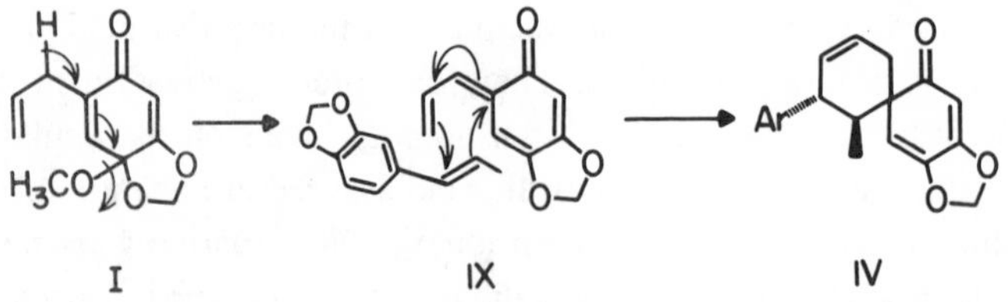

SCHEME 22.3

ity. The defenders of carbenium ion intermediates, however, customarily appeal to the much faster cyclization—such as routes **A** and **B**—than C–C bond rotation to account for the preservation of the configuration of starting materials. Besides, the production of thermodynamically favored trans derivatives and the "wrong" orientation of the concerted coupling—represented by structure **IX**—required to construct product **IV** speak in favor of the carbenium ion hypothesis.

As it turns out, neither position is correct, for an additional piece of evi-

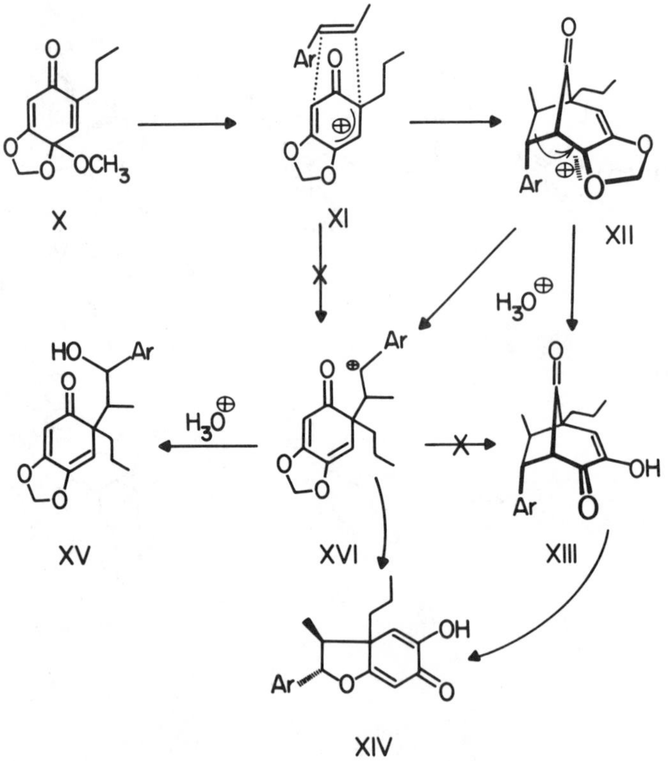

SCHEME 22.4

dence supports a third mechanism. When *trans*-isosafrole (**II**) and ketal (**X**)—in which the allyl chain has been replaced by an unreactive propyl chain—are brought into contact under *p*-toluenesulfonic acid catalysis in a polar solvent (acetonitrile), adducts **XIII, XIV,** and **XV** are produced (see Scheme 22.4). While the carbinol **XV** would apparently support the participation of a **VI**-type benzylic cation, the production of thermodynamically unfavored *endo*-aryl bicyclo[3.2.1]heptane derivative **XIII** would be completely inconsistent with cationic open chain intermediates. In fact, the concerted [3+2] cycloaddition indicated in **XI** has been postulated to account for the construction of the bicyclic ketone.[1]

It has also been shown that **XIII** is converted into **XIV**[1] and vice versa[2] upon contact with triflic acid in acetonitrile, thus strongly suggesting the intermediacy of the bicyclic structure in the route that leads to **XIV** and **III.** Its intermediate precursor **XII** would also be amenable to fragmentation to give carbocation **XVI** that obviously precedes carbinol **XV.** Finally, the slow conversion of the latter to **XIV** under *p*-toluenesulfonic acid catalysis has also been observed.

REFERENCES

2. M. A. De Alvarenga, U. Brocksom, O. R. Gottlieb, and M. Yoshida, *J. Chem. Soc. Chem. Commun.*, 831 (1978).

PROBLEM 23

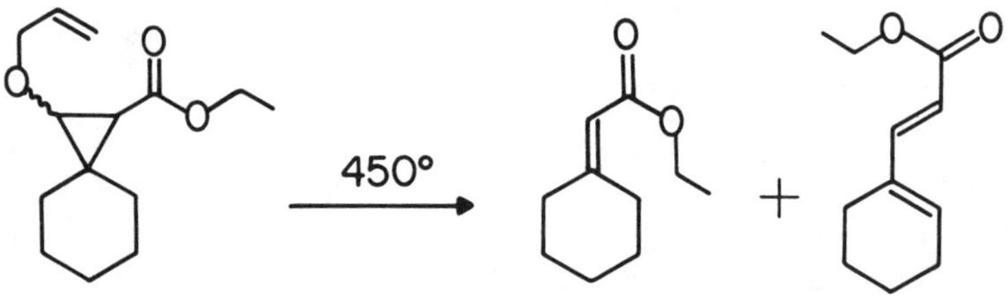

450°

I II III

1. E. Wenkert and J. R. de Sousa, *Synth. Commun.*, **7**, 457 (1977).

PROBLEM 23
Nonselective Unraveling of Cyclopropane

That cyclopropane is an unstable entity at moderately high temperatures has been recognized for many years. Yet the introduction of polar substituents modifies the thermal and acid catalyzed behavior of the trimethylene so substantially that this versatile system is still open to surprising chemistry (see also Problem 3).

The reactivity pattern exhibited by β-oxycyclopropyl-α-carbonyl compounds revolves chiefly around the selective fragmentation of the C–C bond situated between the oxy and carbonyl substituents, by virtue of the possible transmission of electron density between donor and electron units.[2] Well defined polar intermediates, that is, zwitterionic species, are likely to intervene in some cyclopropyl–carbonyl transfigurations.

Along these lines, one possible decomposition route that would account for product **II** would include one such selective heterolytic unraveling to give **IV.** Catalysis by active spots in the glass surface of glassware might well provide some stabilizing assistance since the pyrolysis reactor apparently was not pretreated with neutralizing agents such as chlorotrimethylsilane or amines.

At this point, participation of the allyl side chain must be recognized, because an *n*-butoxy derivative reacts in a different fashion.[1] At least two routes, **A** and **B,** are open to pivotal intermediate **IV** (see Scheme 23.1). While the electron reorganization that leads to alkoxycarbene **V** (route **A**) and dihydrofuran (**VII**) via its addition to the terminal alkene[3] would be justified by the high temperatures employed at which other carbenes have been thermally generated, the prior cyclization of zwitterion **IX** (route **B**) and its ensuing 1,2-hydrogen shift followed by C–C bond fragmentation in **X**—the natural precursor of products **II** and **VII**—would perhaps be more feasible.

The over 500 K temperature at which this reaction was performed makes the intermediacy of free radicals appealing. Although the interested reader may easily translate Scheme 23.1 into carbon radical terms, new mechanistic alternatives that account for product **II** may also be postulated that include the intervention of these highly reactive species.

If one takes the unusual step of breaking a different C–C bond of cyclopropane **I** at the beginning, in particular one that will be broken in the end anyway, the resulting diradical not only will be stabilized by the vicinal oxygen atom,

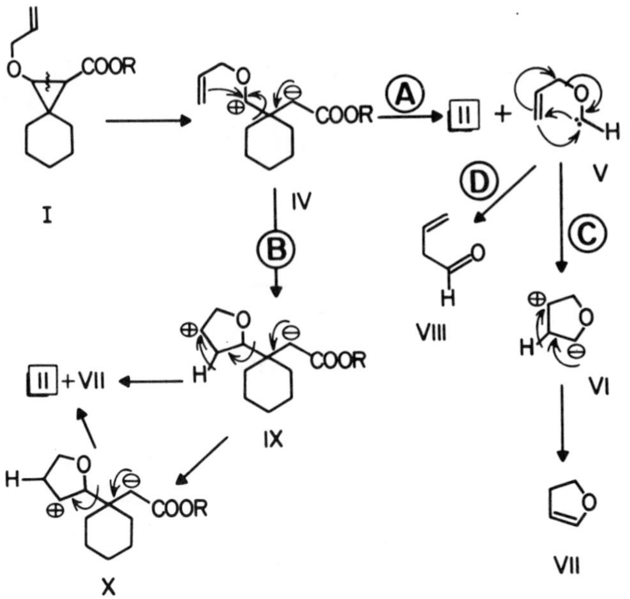

SCHEME 23.1

and a tertiary positioning, but it may also collapse towards compound **II** in the two diverging ways as indicated in Scheme 23.2

A thoughtful and responsible choice among these four options is presently not possible, for on one hand the nature of the by-products that might have given some useful evidence was not ascertained, and on the other, a number of related, systematically modified starting cyclopropanes should be experimentally examined first. At present, these results are not available.

As for compound **III,** it obviously results from the selective disconnection of the third C–C bond of cyclopropane **I,** something that certainly underscores the remarkable versatility of β-oxycyclopropyl ketones and esters. Although the extrusion of allyl alcohol indicated in route **G** of Scheme 23.3 would readily provide an explanation for the formation of dienic ester **III,** a totally different phenomenon could also be involved. The distance between the carbonyl oxygen and its equatorial γ proton on the six-membered ring is short enough (roughly 2.4 Å) to cause attractive interaction between them. The molecular polarization that results would trigger the selective rupture of the cyclopropane ring with simultaneous transfer of this γ proton to the carbonyl oxygen. This process is known as a homo-1,5-hydrogen shift. An electrocyclic reaction equivalent to a

SCHEME 23.2

SCHEME 23.3

second 1,5-hydrogen transfer in the resulting enol ester **XV** would liberate allyl alcohol from this intermediate, leaving behind dienic ester **III.**

The first example of this transformation appeared in the literature as early as 1939.[4] It has been extensively studied ever since in the case of cyclopropane esters and ketones,[5] vinyl cyclopropanes,[6] and vinyl epoxides,[7] a remarkable multiplicator effect of any piece of research indeed.

REFERENCES

2. The β-oxycyclopropane carbonyl systems have been used extensively, among other things, for the synthesis of useful 1,4-dicarbonyl compounds, and cyclopentenones derived therefrom. For leading references see M. Julia and M. Tchernhoff, *Bull. Soc. Chim. Fr.*, 181 and 185 (1956); E. Wenkert, N. F. Golob, R. P. Hatch, D. Wenkert, and R. Pellicciari, *Helv. Chim. Acta*, **60,** 1 (1977); E. Wenkert, *Heterocycles*, **44,** 1703 (1980); E. Wenkert, R. S. Greenberg, and M. S. Raju, *J. Org. Chem.*, **50,** 4681 (1985).
3. Notably, route **A** involves the production of either dihydrofuran (**VII**) or 3-butenal (**VIII**) depending on whether **V** acts as an electrophilic carbene (route **C**) or a nucleophilic unit (route **D**) as one would expect from its being vicinal to an allyloxy unit.
4. C. D. Hunt and M. A. Pollack, *J. Org. Chem.*, **3,** 550 (1939).
5. J. J. Gajewski, R. J. Weber, R. Braun, M. L. Manion, and B. Hymen, *J. Am. Chem. Soc.*, **99,** 816 (1977) and references cited therein.
6. G. D. Andrews and J. E. Baldwin, *J. Am. Chem. Soc.*, **98,** 6705 (1976) and references cited therein.
7. J.-C. Paladini and J. Chuche, *Bull. Soc. Chim. Fr.*, 197 (1974) and references cited therein.

PROBLEM 24

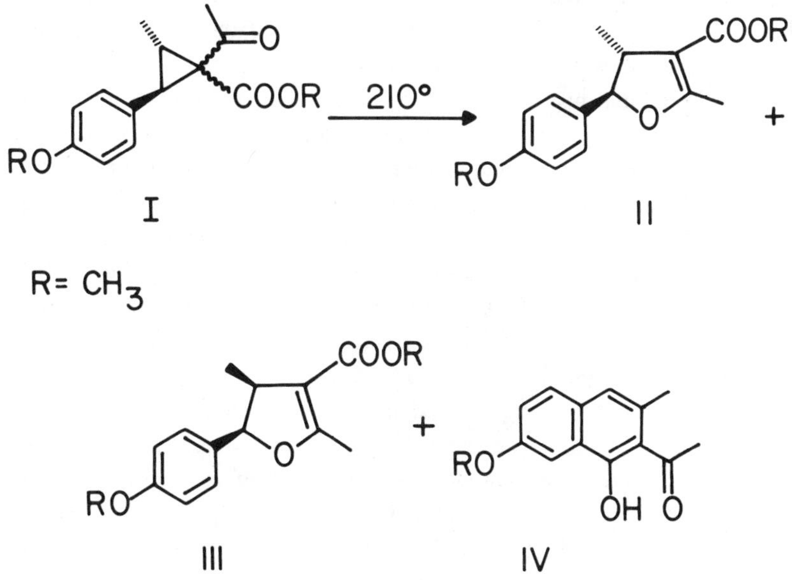

I → (210°) II + III + IV

R= CH_3

1. M. E. Alonso, W. Chitty, M. de L. Borgo, and S. Pekerar, *J. Chem. Soc. Chem. Commun.*, 1452 (1984).

PROBLEM 24
Selective Unraveling of Cyclopropane

This experiment, which combines well anticipated and totally unexpected results, was designed to compare the thermal versus the Lewis acid-catalyzed rearrangement of a doubly activated cyclopropane. This twofold activation was understood in terms of the electron donating effect of the *p*-anisyl substituent at C-2 that is complementary to the familiar electron withdrawal of the 1,1-diacyl substitution, which closely resembles Problems 15 and 23. This combination, called captodative cyclopropanes, considerably facilitates some of the various ring unraveling pathways open to the trimethylene ring.

It was found that just contact with neutral alumina at room temperature in benzene or chloroform suffices to promote quantitative isomerization of compound **I** to the *trans*-dihydrofuran derivative **II** only, with retention of the configuration of the aryl and methyl substituents of cyclopropane **I**.[2] This highly unusual result was underscored by the formation of both **II** and **III** during the parallel thermolysis experiment, a result that follows well established lines. Thus, dihydrofurans **II** and **III** are very probably the result of a 1,3-sigmatropic rearrangement formally similar to that described in Problem 14.

However, compound **IV** was also present in the reaction mixture in as much as 27%. Its formation was unpredicted and unprecedented. Comparison of empirical formulas of **I** and **IV** indicates that a fragment composed of one carbon, four hydrogens, and an oxygen atom, the elements of methanol, is separated from the starting material, in all probability from the ester function. Although dissection of naphthol **IV** (Scheme 24.1) does not provide an obvious connection between the composing fragments and the starting materials, the β-keto

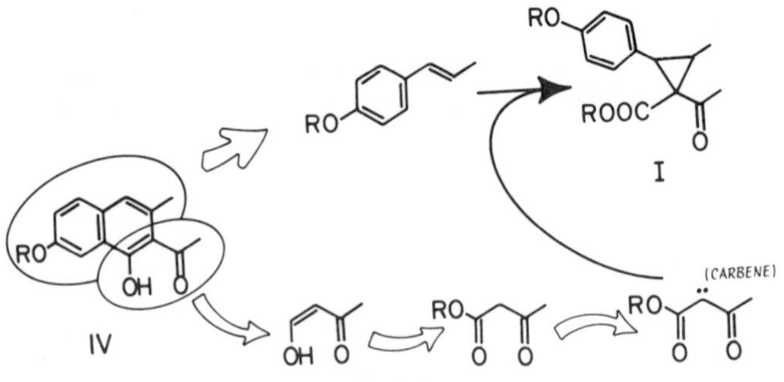

SCHEME 24.1

ester portion of **I** becomes apparent. In addition, the propenyl-*p*-anisole fragment, actually the naturally occurring *trans*-anethole, can be traced to the C_2–C_3 portion of the starting cyclopropyl ester **I.** Finally, the functionalization of the ortho carbon of the aromatic ring that gives rise to the fused ring structure, probably by way of delocalization of a charge, somehow developed at the benzylic position, becomes apparent.

The captodative substitution of **I** could in principle suggest an initial heterolytic cleavage.[3] This is a particularly attractive start since it would readily account for the cyclization step that furnishes the dihydrofurans of mixed stereochemistry. However, there is no plausible way by which the delocalized benzylic cation could become bonded to the ester carbonyl carbon with displacement of methanol. While the alternative ketene structure **VI** might well be derived from the key intermediate **V,** its cyclization to **VII** would be precluded by an unprecedented nucleophilic attack of the electrophilic central carbon of the ketene moiety on the benzylic carbenium ion (see pathway **A** of Scheme 24.2) and the great instability expected for the remaining cation flanked by two keto functions in **VII.** An ad hoc proton elimination indicated in **VI** would provide a way out (route **B**). This option nevertheless would be unlikely since a regular thermodynamically controlled elimination in the opposite direction to give the styrene derivative **IX** would be far more favorable. But in that case, it is conceivable that a thermal electrocyclic reaction of **IX** involving six π electrons from which diketone **VIII** would result (see option **C**).

Alternatively, although the reaction temperature is rather moderate, the possibility of a homolytic cyclopropane ring cleavage must also be considered.

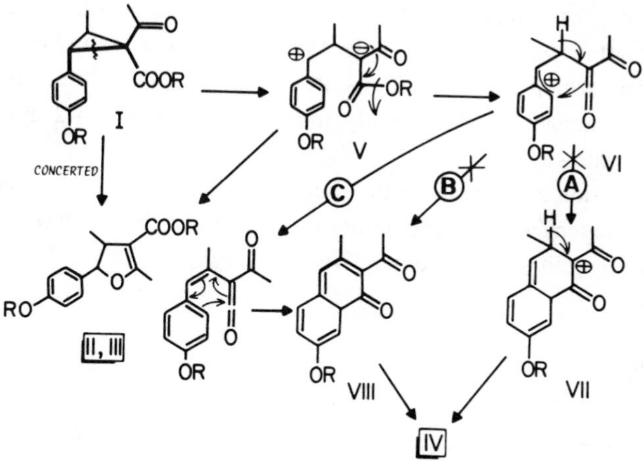

SCHEME 24.2

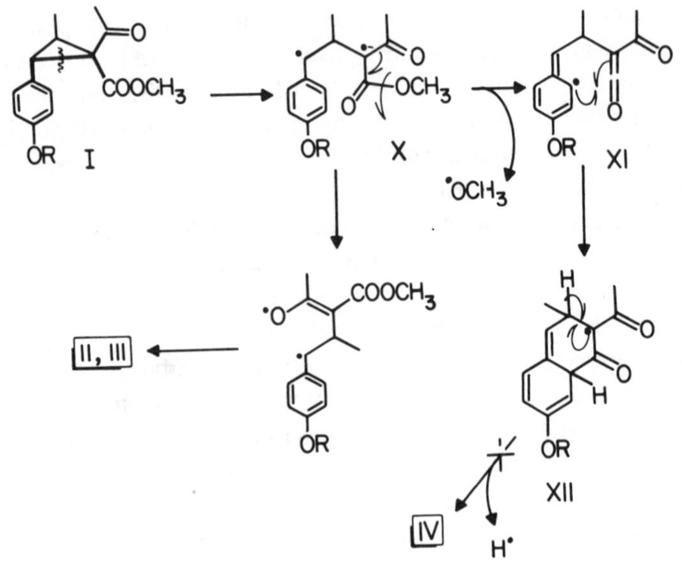

SCHEME 24.3

If delocalization and coupling of diradical **X** yields dihydrofurans **II** and **III** after rotational equilibration, conceivably it might undergo alternative elimination of a methoxyl radical to give ketene derivative **XI** (see Scheme 24.3). Its further cyclization to **XII** and the subsequent steps of extrusion of hydrogen and enolization would be driven by the aromatization energy gained through this sequence. However, methoxy radicals are eliminated only when a very strong driving force such as peroxide decomposition exists. Such reaction conditions are not present in this example. In addition, there are no known reactions in solution in which hydrogen atoms are eliminated. Therefore, the radical alternative is untenable. Consequently, route **C** of Scheme 24.2 appears as the most likely pathway.

This reaction appears to be the first of a new general class, since replacement of hydrogen, methyl, and chloro substituents by methoxy on the aromatic ring of **I** also yield naphthol derivatives.

REFERENCES

2. M. E. Alonso and A. Morales, *J. Org. Chem.*, **45,** 4530 (1980).
3. A. B. Chmurny and D. J. Cram, *J. Am. Chem. Soc.*, **95,** 4237 (1973).

PROBLEM 25

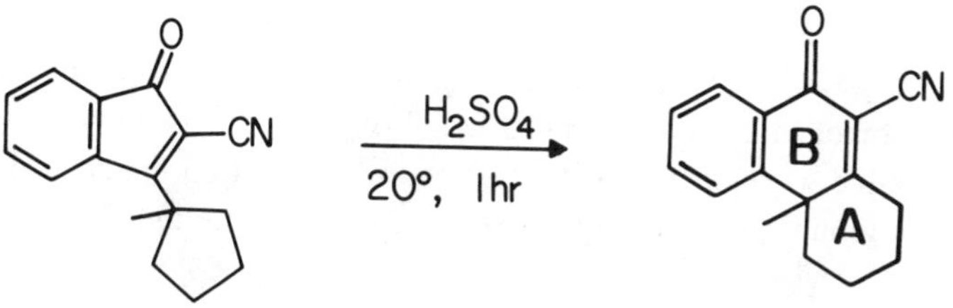

H$_2$SO$_4$
20°, 1 hr

I

II (65%)

1. E. Campaigne and R. A. Forsch, *J. Org. Chem.*, **43,** 1044 (1978).

PROBLEM 25
Multiple Alkyl Scrambling with Ring Expansion

This is an unusual case of two vicinal five-membered rings being converted into two fused six-membered rings, without further alterations of other functional groups on the molecule, notably nitrile. This means that water must have been rigorously excluded to prevent the hydrolysis of this function to amide or carboxylic acid, since the reaction medium is strongly acidic.

The heterolytic disconnection of the benzyl–vinyl C–C bond is an obvious requisite for the construction of ring **B** of target compound **II.** However, this operation cannot be performed directly owing to the great instability of the resulting charged species. Modification of the vinyl carbon β to the ketone appears to be, therefore, an indispensible condition before any other transformation can take place.

One way to achieve this would be by means of the acid promoted polarization of compound **I.** The proton–carbonyl oxygen association would reduce considerably the π-electron density of the already electron deficient β carbon of the enone group, probably to the extent of a carbenium ion (**III**) in sulfuric acid.

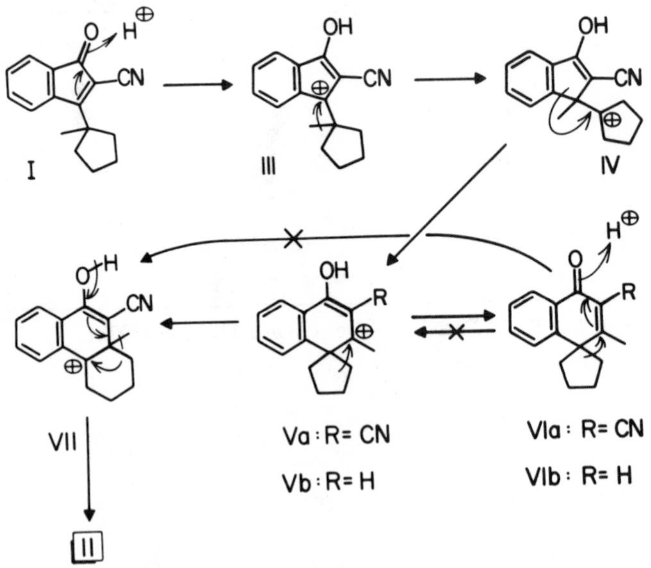

I III IV VII Va : R= CN Vb : R= H VIa : R= CN VIb : R= H II

SCHEME 25.1

The positioning of a quaternary center directly bonded to this cation makes feasible, on one hand, 1,2-migration of the quaternary methyl following Wagner–Meerwein rearrangement lines, the consequence of which would be **IV.** This intermediate, in turn, contains a similar combination of a tertiary carbenium ion in the vicinity of another quaternary center. This circumstance would now be favorable for the disconnection of the five-membered ring of the indene system, hence permitting the construction of the final six-membered ring, as indicated in **VIa** (see Scheme 25-1).

Ring B of compound **II** might be formed from intermediate **Va** by ring expansion although the angular methyl would end up on the wrong side of the molecule. Nevertheless, the 1,2-migration of this methyl to the benzilic cation next to it (**VII**) furnishes the right configuration.

Conceivably, the spirocyclopentane **VIa** should be an almost inevitable by-product of this sequence. Its presence would be diagnostic of the validity of the proposal depicted in Scheme 25.1. However, **VIa** could not be detected in the reaction medium. Nonetheless, a similar compound, **VIb,** was independently synthesized and exposed to strong mineral acid on the assumption that it should yield cation **Vb** nearly as easily as the starting enone **I** yields **III.** However, indenone **VIb** proved to be stable in 96% sulfuric acid.

It could be argued that compound **VIb** is not a good model of **VIa** since those unsaturations in both compounds from which the pivotal carbenium ion is to be derived are basically different. However, the additional experimental fact that **VIII** rearranges to **IX** without interference from methyl migration [**X** was not detected, see reference (2)], casts further doubts on the validity of this sequence (see Scheme 25.2).[3]

Failures are frequently as instructive as successes in science. The experi-

VIII

$HClO_4$

IX (100%)

X

SCHEME 25.2

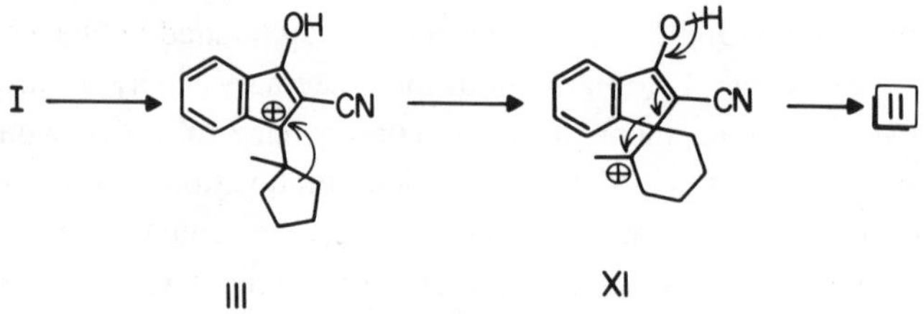

SCHEME 25.3

ence just described suggests the operation of an alternative migration sequence that would include a shift of one of the cyclopentane C–C bonds in lieu of the methyl 1,2-migration. Indeed, the tertiary cation (**XI**) also provides an adequate basis to break the indene benzyl C–C bond under the auspices of the enol moiety (see Scheme 25.3). This would furnish a more direct route to product **II.**

It is worth noticing that the order of ring enlargement of this last alternative runs in the opposite direction to that of the mechanism suggested first.

REFERENCES

2. M. Nakazaki and K. Yamamoto, *Chem. Lett.*, 1051 (1974).
3. Compound **I** was interpreted as a vinylog of **VIII.**[1] This leaves room to wonder to what extent **i** and **ii** are comparable species.

PROBLEM 26

$$\text{I} \xrightarrow[\text{20°, 30'}]{\text{SOCl}_2 \text{ (EXCESS)}} \text{II (83\%)}$$

I

II (83%)

---

1. K. Senga, M. Ichiba, and S. Nishigaki, *J. Org. Chem.*, **43,** 1677 (1978).

PROBLEM 26
First Encounter with Sulfur—In Situ Construction of Sulfoxide and Its Pummerer Rearrangement

During the first stages of their professional practice many young chemists tend to stay at a prudent distance from heterocyclic compounds probably because their great variety and apparently inextricable reaction patterns leave an overwhelming impression. However, heterocycles not only are fascinating and useful molecules, but they are used in nature more than anything else for the intricate processing of living matter, from chlorophyl and ATP to proteins, DNA, secondary metabolites, and other biomacromolecules.

The previous reaction describes the synthesis of a novel class of heterocycles by the name of [1,2,3]thiadiazolo[4,5-*d*]-pyrimidines. Visibly, a redox process is involved, whereby the initial hydrazine derivative becomes an R–N=N–R′ system and, at the same time, the sulfur atom in thionyl chloride is converted to an azo-sulfide. The transfer of oxidaton level from sulfur—as sulfoxide—to the neighboring atom is a well documented process in sulfur chemistry that is called the Pummerer rearrangement.[2] In essence, it involves the treatment of sulfoxides with an electrophile such as acetic anhydride to yield an α-acetoxy sulfide, according to the following sequence (see Scheme 26.1):

Conceivably, the azo-π system of **II** could be derived from the internal redox reaction implied in the elimination–addition sequence depicted in the production of **III.** Subsequent β elimination would afford the desired double bond. In the case of nitrogen as the neighboring atom the ylide intermediate of Scheme 26.1 would not be necessary since the availability of a nonbonding pair of electrons should make the use of base accessory. Thus the reaction pathway would be subsequently shortened. Moreover, the required electrophile, thionyl chloride, would be present in the reaction medium in large quantity. The prospect of a Pummerer-type rearrangement during the transformation of **I** into **II** is therefore likely.

As for the C–S bond in **II,** it must not pass unnoticed that the fragment to the right of the dotted line in **I** (see Scheme 26.2) may be taken as a vinylogous urethane that could well serve as receptor of an enamine-type attack of some electrophilic form of sulfur. This sulfur group, in turn, would result from an initial interaction of the thermal nitrogen onto thionyl chloride as depicted in **IV** and **V.** The remaining steps of this scheme would be understood as a vinylogous version of the Pummerer rearrangement.

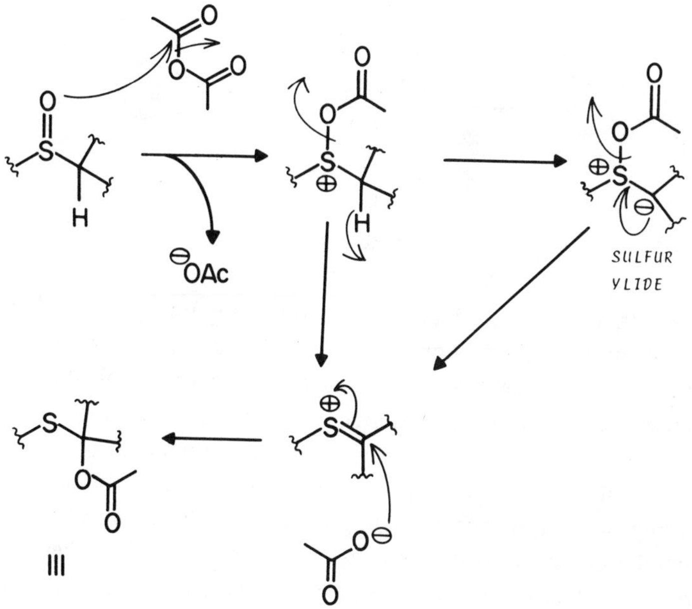

SULFUR YLIDE

III

SCHEME 26.1

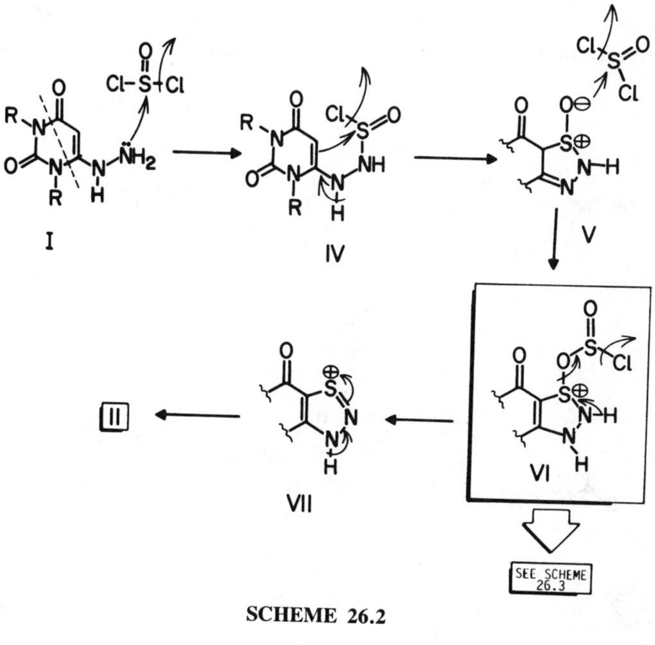

I

IV

V

VI

VII

II

SEE SCHEME 26.3

SCHEME 26.2

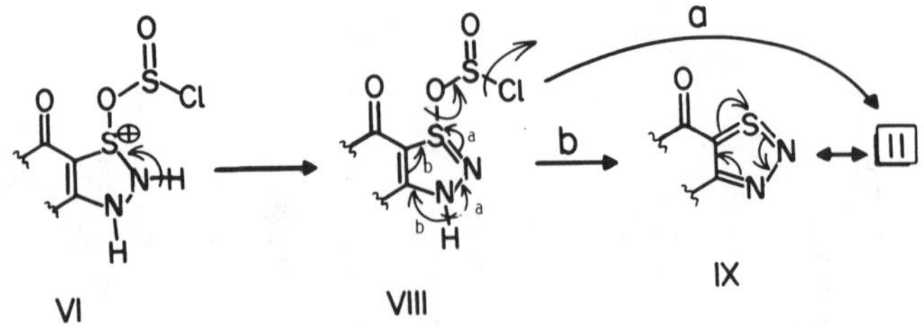

SCHEME 26.3

The authors of this work proposed a somewhat different evolution of intermediate **VI** with the production of structure **IX** (see Scheme 26.3). This structure, geometrically unacceptable for a carbon cumulene, is allowed for sulfur.

Subtle changes in molecular structures often trigger remarkably different reaction routes and the present example is a case in point. If one of the methyl substituents in compound **I** is replaced by a proton, that is, compound **X** is now the starting material of our problem, an entirely different mechanism sets in, although the end result is the same as the one described in Scheme 26.3 (see Scheme 26.4). This differentiation was made possible through the isolation and characterization of compound **XII.** This material could be converted into thiadiazole **XV** by treatment with thionyl chloride in hot ethanol. Although it is quite easy to see how **XII** was formed, it will take a little more imagination to figure out how the novel 1,3-sulfur migration, that is, steps **XII–XIV,** take place. The required S–N bond disconnection, actually the cause of instability

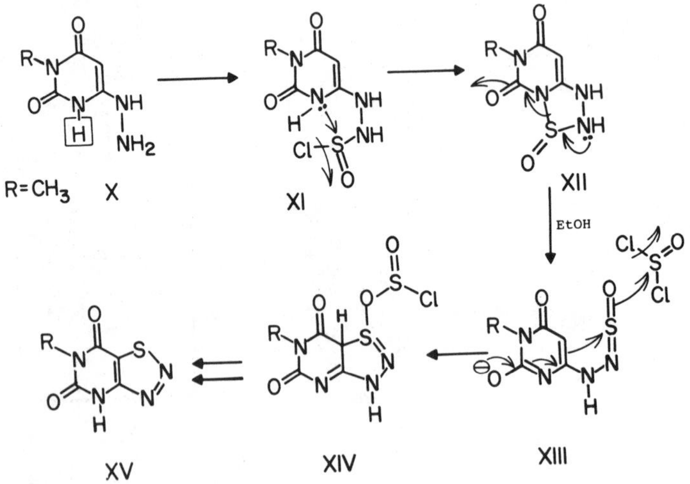

SCHEME 26.4

of **XII,** should be aided by electron withdrawal of the uracyl function. The unusual ketene thioimine **XIII** that would result from this bond breakage would be suitable for nucleophilic attack on the sulfur atom by the now familiar en-amino function, with the ensuing incorporation of thionyl chloride in **XIII.** The reaction pathway thereafter follows the same course described for the synthesis of **II.**

REFERENCES

2. For a review see E. Block, *Reactions of Organosulfur Compounds*, Academic, New York 1978, pp. 154–162.

PROBLEM 27

Cl, N, S, PbOAc$_4$, HOAc, OAc, CHO

I

II

III

IV

1. J. B. Press and N. H. Eudy, *J. Org. Chem.* **49,** 117 (1984).

PROBLEM 27
Circuitous Oxidation with Lead Tetraacetate

Benzene-fused seven- and eight-membered ring heterocycles such as **I** are important compounds in the pharmaceutical industry owing to their effect as cardiovascular agents.[2] Their complex chemistry ought to be, therefore, of some relevance to anyone interested in this broad field of organic chemistry. The present case is an example of the versatile chemistry of these heterocycles.

The reaction conditions employed here are, no doubt, oxidative. Pb(IV) is capable of receiving two electrons from the organic substrate to become Pb(II) as acetate.[3] This coincides with the number of electrons transferred during one-step oxidations, that is, alcohol to aldehyde, aldehyde to ester, and so on. As a consequence, 1 mol of lead tetraacetate (LTA) will be consumed for every oxidative step.

In order to assess the extent of oxidation of compound **I** in its transformation to **II**, let us consider the following points:

1. The ring contraction that is obviously involved in this reaction ought to include C–C bond disconnection and C–C bond reconstruction at a different point. These two operations as a whole are nonoxidative. Therefore, electron transfer must occur at a stage other than ring contraction itself.
2. Compounds **II** and **III** differ only in one C=C bond. Their separation as far as oxidative levels is therefore one step, that is, two-electron oxidation, wherein 1 mol of LTA is consumed.
3. Similarly, compounds **III** and **IV** differ only in the acetate–aldehydo function at the end of the side chain. Again a two-electron transfer must occur with a second mole of LTA being required.
4. The difference in oxidation levels of **I** and **II** is not so obvious and requires detailed analysis. First, it is important to recognize that the liberation of the two carbon side chain with the simultaneous introduction of acetate is in actual fact an oxidation. Skeletal carbon–hydrogen budgets of **I** and **II**—not including the acetate atoms since this is the added function—reveals that **I** is converted to a system (**II**) in which a proton is missing. It is evident then that no more than one two-electron oxidation mediates in this conversion, with the addition of an acetate molecule.

The authors of this work[1] report in their original paper that the use of 1.1 equivalents of LTA yields only **II** and **III** but no **IV,** while the latter appears when two equivalents of the lead oxidant are employed. Moreover, further excesses of LTA make compound **IV** the predominant product. This experience is in agreement with points (2) and (3).

The question then becomes—how does Pb(IV) become associated to **I** for the required electron transfer to occur? Two anchoring places in **I** are conceivable, sulfur and nitrogen atoms, with precedent in the literature. While thiols are oxidized readily to disulfides[4] and sulfides are converted to sulfoxides,[5] aromatic amines are transformed to amides by LTA.[6] However, the nitrogen atom of **I** is not part of an aromatic amine but an imine instead, and little if anything is known about the coupling of this unsaturated function with LTA. Besides, products **II–IV** exhibit this imine function unchanged. Consequently, compound **I,** as it stands, is likely to become linked with lead at the sulfur atom.

Before going any further, let us examine the ring contraction process. It implies two possible pathways:

1. Coupling of the carbon α to sulfur with the electrophilic imine carbon followed by fragmentation of the resulting four-membered ring (see Scheme 27.1).
2. Coupling of sulfur with the carbon α to the imine function with subsequent fragmentation.

The interpretation of these two schematic views confronts some difficulties. Option (1) requires the extrusion of a hydrogen from a methylene hardly active in acidic medium (acetic acid). Under basic conditions, it would be necessary first to convert the sulfide to a sulfoxide such as **V** to achieve sufficient activation of this methylene. This oxidation may be accomplished by LTA as indicated previously. Nevertheless, the ensuing cyclobutane fragmentation in **VI**—the actual oxidative step—by concurrent attack of acetate and departure of Pb(II) diacetate (X = $PbOAc_2$ in Scheme 27.1) has no precedent in LTA–amine chemistry, although it is electronically balanced. Besides, the final product of this sequence would be sulfoxide **VII.** Having no reductive work-up procedure, this sulfoxide should survive until the isolation step. Since this is not the experimental fact, option (1) must then be discarded.

As for option (2), the first step would have to be preceded by an oxidative operation that would introduce a leaving group Y (see structure **VIII**) on the

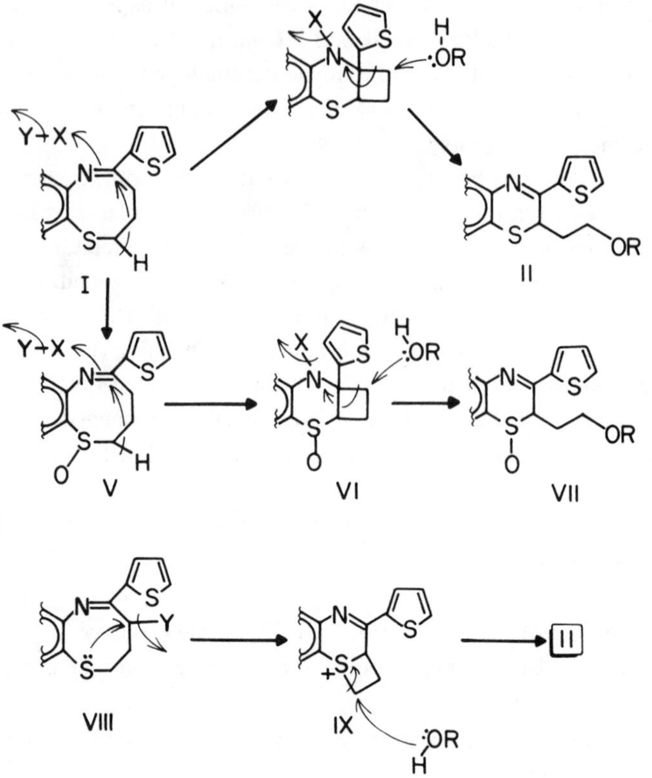

SCHEME 27.1

allylic position. This may be accomplished in principle if one recalls that enolizable ketones are oxidized efficiently to α-acetoxyketones on exposure to LTA, something that is believed to proceed via enol ether **X** (see Scheme 27.2). If imines are looked upon for a moment as nitrogen homologs of ketones, it is also conceivable that these units may be oxidized to α-acetoxyimines by LTA, although to our knowledge, no experimental evidence to support this idea exists in the literature.[7]

Compound **VIII′** thus formed would then undergo ring contraction as postulated before, followed by a now logical cyclobutane ring opening that would occur under the auspices of acetate and the C–C bond weakening effect of the sulfonium unit.

A third mechanism may be postulated if it is assumed that the 1,3-hydrogen shift of the imino group in **I** to its enamino form, which is nucleophilic at the β-vinyl carbon, occurs at an early stage. The electrophilic atom to which that

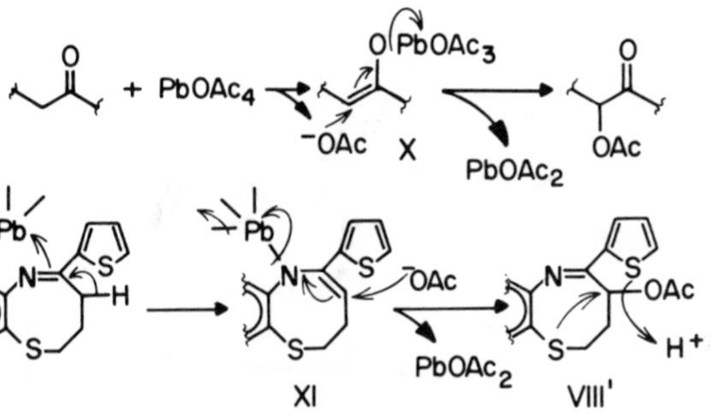

SCHEME 27.2

nucleophilic carbon would attack would be developed by means of association of LTA to sulfur as indicated in **XIII.** Again, the formation of a C–S bond would result in the pivotal cyclobutyl intermediate **IX** whose conversion to **II** has already been commented on (see Scheme 27.3).

Finally, a fourth mechanism would be based on the fact that LTA may be bonded to both nitrogen and sulfur in **XII** with displacement of 2 moles of acetate from lead, to give **XIV.** The sulfonium sector would enhance the lability of its α carbon towards acetate, thus furnishing intermediate **XV.** This would secure the construction of the acetate-bearing side chain, while still holding the lead atom in a pincerlike structure. This molecule would be ideally suited sterically and electronically for the favorable six-electron electrocyclic process depicted in **XV.** This electron reorganization would yield product **II** directly.

Undoubtedly, the combination LTA–compound **I** is open to a number of

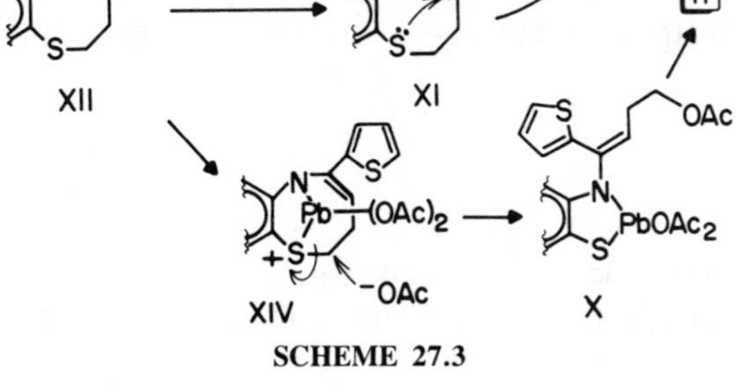

SCHEME 27.3

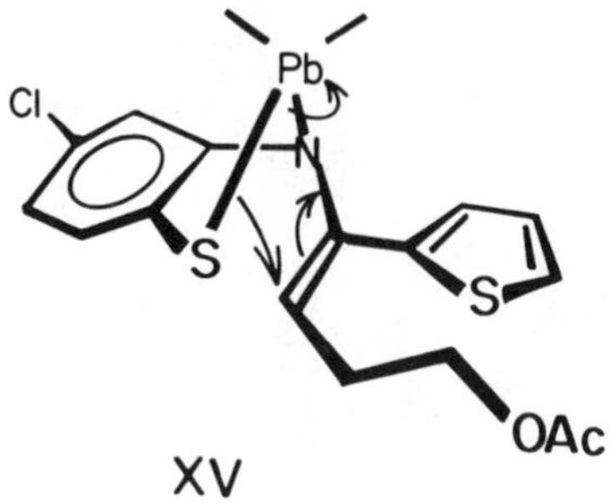

XV

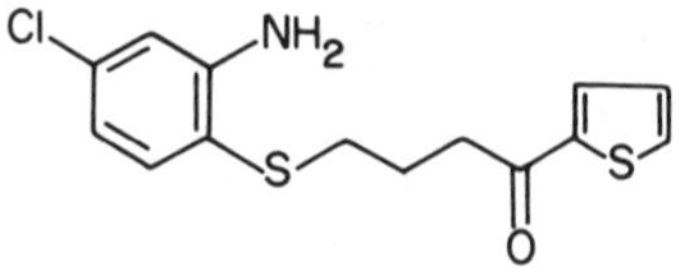

XVI

pathways. Of these, only options (1) and (2) may be discarded owing to the lack of precedent of imine α-acetoxylation reactions (but there is always a first time!). The other hypothetical routes appear to rest on solid enough bases and a final choice is difficult. Two more experimental facts[1] add interesting information. The oxygen analog of **I** does not undergo transformation to the corresponding analogs of **II–IV.** This clearly indicates that the sulfur atom is a requisite for ring contraction. Nevertheless, this information still is not helpful for mechanistic choice because it allows for the operation of all routes proposed here. Furthermore, mercuric diacetate does not induce ring contraction of **I,** as opposed to what could be expected, owing to the frequently parallel behavior with that of LTA. Rather, it transforms **I** into amino ketone **XVI.** This reaction is in itself an interesting problem the reader is urged to solve, and probably suggests that the double coordination of the metal atom, more likely to occur in Pb(IV) than in Hg(II) in **XIV** is essential for ring contraction.

REFERENCES

2. They include valium and several other diazepams. See, for example, H. Kugita, H. Inoue, M. Ikezaki, M. Konda, and S. Takeo, *Chem. Pharm. Bull.*, **19,** 595 (1971) and references cited therein.

3. For a recent review on LTA chemistry see G. M. Rubottom, *Oxidations in Organic Chemistry*, W. S. Trahanowsky, Ed., Academic, New York, 1982, Part D.
4. L. Field and J. E. Lawson, *J. Am. Chem. Soc.*, **80,** 838 (1958).
5. Oxidation of sulfides to sulfoxides, however, is more efficiently accomplished by reaction with sodium metaperiodate, or hydrogen peroxide, ozone, peracids, manganese dioxide, nitric acid, chromic acid, and other oxidants. See, for example, N. J. Leonard and C. R. Johnson, *J. Org. Chem.*, **27,** 283 (1962).
6. L. Horner, E. Winkelmann, K. H. Knapp, and W. Ludwig, *Chem. Ber.*, **92,** 288 (1959).
7. That imines are enolizable systems is well established. Enamine synthesis is in part based on this principle.

PROBLEM 28

I + II $\xrightarrow[60°]{CH_3CN\text{-}THF}$ III (38%)

1. R. Okazaki, F. Ishii, K. Sunagawa, and N. Inamoto, *Chem. Lett.*, 51 (1978).

PROBLEM 28
Second Encounter with Sulfur—Interaction of an Enamine and a Dithioketene Acetal

Sulfur compounds are exceedingly interesting and useful systems. Their impressive versatility is the result of their complex chemistry, which any self-respecting organic chemist should be acquainted with.

Although we have brushed lightly over some sulfur-related transformations in past problems, the next two exercises touch upon more central aspects of organosulfur chemistry that provide challenging mechanisms.

Inspection of empirical formulas and the corresponding structures of starting material and product reveals that **III** is solely a dimer of **I.** Two important facts, however, conspire against this outer-layer simplicity. First, if the deep-blue compound **I** is allowed to dimerize by itself (and this in fact occurs in solution), the colorless dimeric structure **IV** (but not **III**) results. Second, there is an enamine in the medium, which obviously makes the difference. The role of this enamine should not be taken lightly, particularly if the conjugation of the sulfur atoms in **I** imbues this compound with a behavior reminiscent of oxygen homolog systems, that is, quinonoids.[2] The exocyclic double bond is amenable to a Michael-type 1,4 addition by suitable nucleophiles while the thioketone sulfur is potentially a nucleophilic center (see Scheme 28.1).

This quality makes compound **I** labile towards a moderate carbon nucleophile such as that of an enamine. From the interaction of these two species,

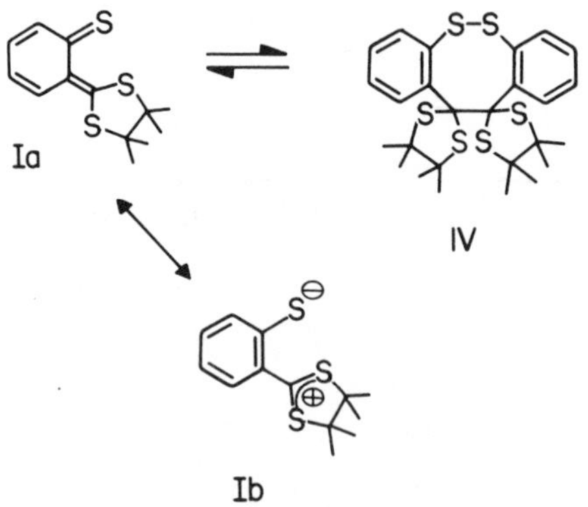

Ia IV Ib

SCHEME 28.1

zwitterion **V** would result (see Scheme 28.2). One obvious outlet to this sequence would be the termination of the [2+4] cycloaddition to give **VI**, a process that mimics the reaction of **I** with other enamines simpler than **II.**[3]

An equilibrium condition between **V** and **VI,** however, is also conceivable in the present case, owing to the thermodynamic instability implied in the geminal substitution of isopropyl and dimethylamino groupings. The persistence of **VI** in the reaction medium would give opportunity for a bimolecular interaction in the Michael mode with a second mole of substrate. The ensuing rare fragmentation that liberates the amine portion from **VII** looks more familiar if one draws it in parallel with the hydrolysis–decarboxylation of a β-keto ester.

This behavior of the RS–C=C fragment may be inverted by properly handling sulfur polarity. Thus, its oxidation to a vinyl sulfoxide converts this unit into an excellent "two-carbon Michael acceptor"[4]—notice that methyl vinyl ketone and its many equivalents are all three-carbon acceptors. This ambivalent polarizability by simple synthetic operations makes organosulfur synthons very interesting.

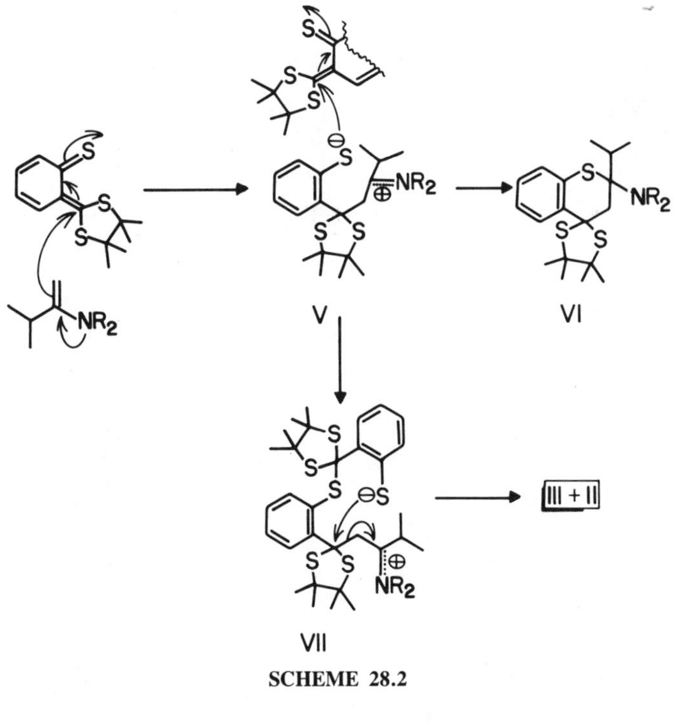

SCHEME 28.2

REFERENCES

2. For a review, see R. Gompper, *Angew. Chem. Int. Ed. Engl.*, **8,** 312 (1969).
3. R. Okazaki, F. Ishii, and N. Inamoto, *Bull. Chem. Soc. Jpn.*, **51,** 2 (1978); I. Fleming and M. H. Karger, *J. Chem. Soc.*, (*C*), 226 (1967). See also R. Okazaki, K.-T. Kang, K. Sunagawa, and N. Inamoto, *Chem. Lett.*, 55 (1978).
4. J. Schlessinger, *Tetrahedron Lett.*, 4711 and 4755 (1973); T. Oishi, *Tetrahedron Lett.*, 3757 (1974).

PROBLEM 29

I + II $\xrightarrow{Et_3N,\ DMF}$ III (not formed)

I + II $\xrightarrow{Et_3N,\ DMF}$ IV (30%)

1. M. Mizuno and M. P. Cava, *J. Org. Chem.*, **43,** 416 (1978).

PROBLEM 29
Third Encounter with Sulfur—Chlorine versus Sulfur as Carbanion Stabilizer

The results described in this problem are actually derived from the reinvestigation of an old subject: The synthesis of compound **IV**, which was first obtained back in 1926.[2] No one showed much interest in this area for 50 years thereafter, but the advent of *organic metals*, among which benzotetrathiafulvalenes are prominent members, brought about the resurrection of this ancient piece of work. A remarkable chemistry was unveiled, including the notable fact that **I** and **II** would not yield the expected mixed thiafulvalene **III.** Compound **IV** was isolated instead.

The formation of **III** and **IV** can both be understood as a series of addition–elimination reactions occurring on the central C=C bond of **I,** by the works of sulfur anions liberated by the interaction of *o*-benzodithiol (**II**) and trimethylamine. However, the direction of the addition reaction cannot be ascertained at first glance, because some confusion may occur if the relative stability of the carbanions resulting from this operation is not carefully analyzed first. For this purpose, there are the handy Baldwin's Rules,[3] although these laws face some problems when applied to second-row elements such as sulfur. Therefore, they will be disregarded here.

On the other hand, there is the well-known fact that sulfur is an excellent carbanion stabilizer. But not for those reasons popularly ascribed to $d\pi$–$p\pi$ backbonding of the carbanion lone pair into the vacant 3*d* orbital of sulfur, as suggested earlier. Recent molecular orbital calculations appear to challenge the validity of this model. These calculations, for example, predict the order of gas phase (no solvent effects to worry about) carbanion stabilization as sulfur > oxygen > carbon whether or not the 3*d* orbitals of sulfur participate in the calculations. Besides, these theoretical considerations surprisingly show that the C–S bond in the thiomethylene carbanion is longer than in the neutral thiomethane parent. Until recently, the prevailing mechanism of stabilization was centered around the classical concept of polarizability of the electron "cloud" around the sulfur nucleus.[4]

The relative electronegativity of heteroatoms directly bonded to the carbanion also contribute to the overall stabilization. This is best illustrated by the reaction of *o*-benzodithiol (**II**) with tetrachloroethylene. Of the two possible cycloaddition routes, that leading to the thermodynamically less stable five-membered heterocycle prevails, perhaps owing to the greater stability of carbanion **VI** as compared with anion **VIII** (see Scheme 29.1). Therefore, it is important to realize that: Chlorine being more electronegative than sulfur lowers the energy level of an adjacent negatively charged carbon more effectively than does sulfur.

This being the case, the direction of the initial addition reaction of **II** to **I** becomes clear. The construction of the predicted product **III** requires the less stable carbanion **X** (see Scheme 29.2) while the preferred dichlorocarbanion **XIII** leads to the observed product **IV.**

It is also interesting to mention that in order to arrive at **IV** the direction of the addition reaction on the C=C bond must be inverted at some point. If the more stable carbanion **XVII** is formed from the key intermediate **XVI,** the sequence of events that follows leads inevitably back to this key intermediate (**XVI**). By contrast, the less stable carbanion **XIX,** which is equivalent to the also unfavored **X,** is the only way to reach the target product **IV.** This less

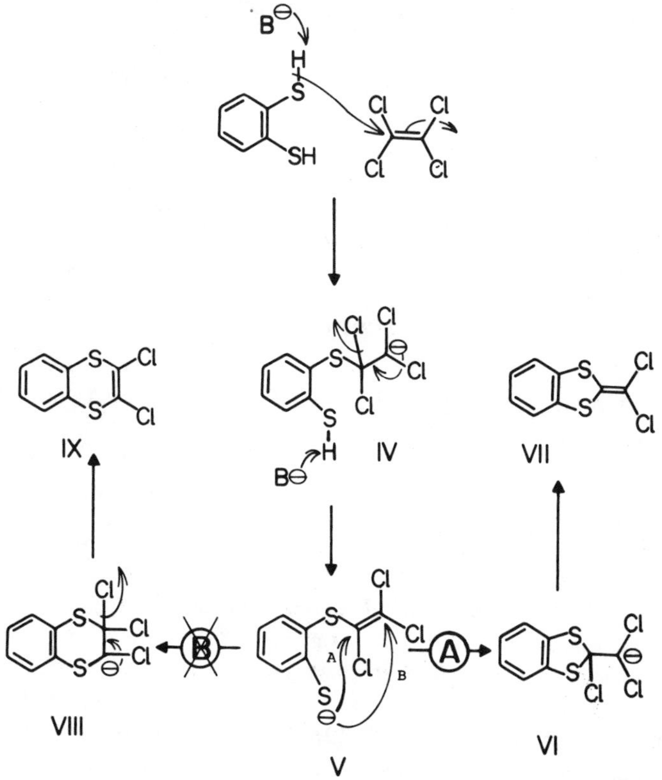

SCHEME 29.1

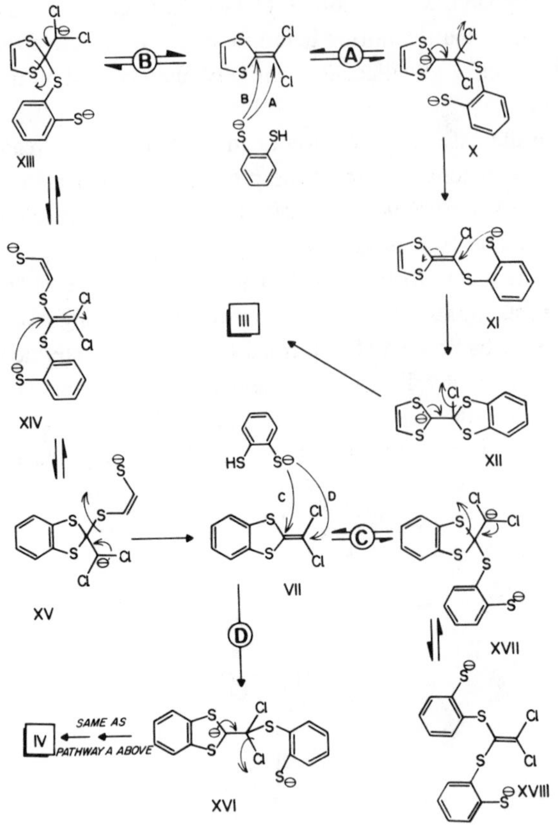

SCHEME 29.2

competitive route is made possible only because the favored pathway is a no-exit vicious circle!

REFERENCES

2. W. R. H. Hurtley and S. Smiles, *J. Chem. Soc.*, 2263 (1926).
3. J. E. Baldwin, *J. Chem. Soc. Chem. Commun.*, 734 (1976).
4. For leading references see D. Seebach, *Angew. Chem. Int. Ed. Engl.*, **8,** 639 (1969); F. G. Bordwell, *J. Org. Chem.*, **42,** 326 (1977); A. Streitweiser, Jr., and J. E. Williams, *J. Am. Chem. Soc.*, **91,** 191 (1975); J.-M. Lehn and G. Wipff, *J. Am. Chem. Soc.*, **98,** 7498 (1976) and references cited therein.

PROBLEM 30

Cl—C(CH$_3$)$_2$—C≡CH + CH$_3$CH$_2$MgI $\xrightarrow[\text{20°, 24 hr}]{\text{Et}_2\text{O}}$

I II

III (32%) + IV (6%) + V (10%) +

VI (16%) + VII (36%)

1. D. J. Pasto, R. H. Shults, J. A. McGrath, and A. Waterhouse, *J. Org. Chem.*, **43,** 1382 (1978).

PROBLEM 30
Multifaceted Performance of a Grignard Reagent

The apparent simplicity and straightforwardness of the classic Grignard reaction makes everyone feel at home when faced with RMgX reagents. One easily forgets, however, that yields rarely go beyond 85% and not much attention is customarily paid to the associated tar, whose sole presence is an indication of competitive complex and sometimes calamitous reaction processes. At times these side reactions become higher yielding than the expected reaction. The present example is such a case. The authors of this work began their paper with the following comment: "The history of propargyl derivatives with Grignard reagents is one of confusion, widely divergent results being reported by various authors."[1,2]

Before advancing any mechanistic interpretation, it is necessary to add some further experimental evidence:

1. Inconsistent results were observed with different batches of magnesium turnings. This fact speaks in favor of other metal impurities present in the magnesium metal employed and inadequate processing of propargyl halide.
2. The product ratio, particularly in reference to products **III, IV,** and **V,** remained the same regardless of the quenching procedure: water, 5% sulfuric acid, saturated ammonium chloride, and so on. This means that products **IV** and **V** are not the consequence of allene isomerization during work-up. Further, dienes **VI** and **VII** are apparent isomerization products of **III,** although without the actual control experiment at hand, this assertion cannot be proved.
3. There is an incomplete utilization of the two molar equivalents of the Grignard reagent. This implies that a newly formed intermediate Grignard species that uses up the first equivalent of magnesium is competing against **II** for the same Grignard reagent consuming process.
4. The reaction of propargyl bromide with a Grignard reagent gave only the corresponding allene, with very little alkynes present. If we accept that bromide is generally a better leaving group than chloride, this evidence points toward an S_N2' displacement to account at least for compound **III.**

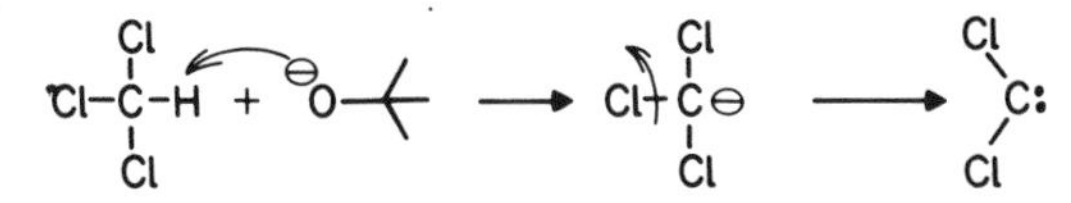

SCHEME 30.1

A different mechanism may be operating, however. This would make use of the relative acidity of the acetylenic proton ($pK_a = 25$). In fact, besides being powerful nucleophiles, Grignard reagents are strong bases and the initial abstraction of the terminal proton of **I** is in order. The resulting acetylide **VIII** would first evolve into a dead end by its association with magnesium to give **IX** and from there back to the starting material (see Scheme 30.1). Conversely, **IX** could undergo a very unlikely bimolecular attack by a second equivalent of **II** to furnish, after work-up, the target terminal alkyne **V.**

An alternative route exists for **VIII,** however. A carbanion with a good leaving group on it may form a carbene upon departure of this leaving group. An example of this is the base treatment of chloroform, illustrated in Scheme 30.2.

If the leaving group is on a carbon other than the carbanion the formation of a carbene is still possible by way of electron delocalization. The best example of this is provided by the progress of intermediate **VIII** (see Scheme 30.3). Electrons in zwitterion **X** that may be derived from **VIII,** may be rearranged to give allenic carbene **XI.** This is nothing but a resonance structure of **X.** This trigonal carbene would then undergo insertion in the C–Mg bond of a second equivalent of **II** to give **XIII.** This is an allenyl Grignard reagent that would be a direct precursor of **III** on one hand. On the other hand, the equilibration of

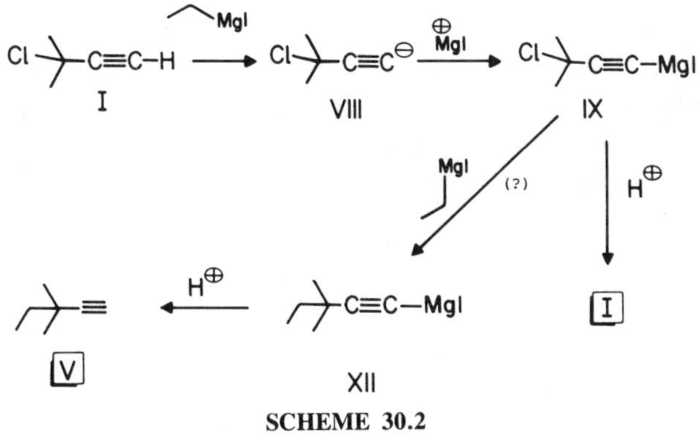

SCHEME 30.2

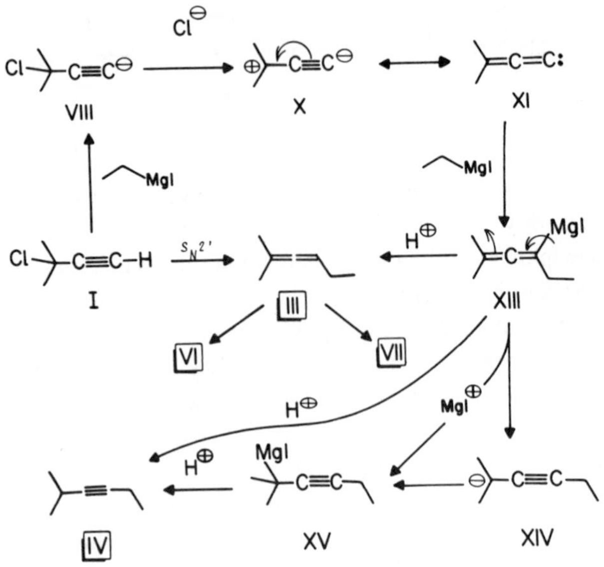

SCHEME 30.3

XIII with propargylic magnesium derivative **XV** would be a facile entry into the otherwise unexplainable **IV.** In addition, dienes **VI** and **VII,** which are probably not the consequence of this scheme, may result from the isomerization of **III** or by means of another rather obscure process not open to this writer's mind.

While it is intellectually pleasing to find self-consistencies in this mechanism, such as the fact that **XII, XIII,** and **XV** would be the Grignard species formed anew that compete against **II** for proton abstraction in the first step (see experimental evidence, (2)), and that the allenic carbene approach is also attractive, there are still other alternative interpretations to be considered. First, the S_N2' pathway that accounts for **II** is supported by experimental evidence (3). The authors of the original paper[3] contend, however, that this route is disfavored in the light of their extended studies about the transition metal salt catalyzed Grignard reactions of propargyl halides, wherefrom allenes invariably result.

Also, this scheme can be easily restructured using free radical intermediates on the grounds that a single electron transfer occurs instead of the more familiar two-electron transfer described here. This option may also be disregarded, however, if one takes into consideration the exclusive formation of alkynes from the free radical chlorination of propyne and allene.[4]

It is worth mentioning that the synthesis of allenes from propargyl halides in Grignard reactions is favored by transition metal catalysis more than by the participation of allenic carbene intermediates in as much as the level of metal impurities in magnesium turnings is not reduced to a minimum, or metal salts such as ferric chloride and other metal chelates such as nickel acetoacetonate are added to the reaction mixture.[5]

REFERENCES

2. The interested reader is invited to follow a most amusing and instructive series of contradictory reports. (a) F. Serratosa, *Tetrahedron Lett.*, 895 (1964); (b) Y. Pasternak and M. Delepine, *C. R. Acad. Sci. Paris*, 3429 (1962); (c) F. Coulomb-Delbecq and J. Gore, *Bull. Soc. Chim. Fr.*, 533 and 541 (1976); T. L. Jacobs and R. A. Meyers, *J. Am. Chem. Soc.*, **86,** 5244 (1964) and **89,** 6177 (1967); (d) D. R. Taylor, *Chem. Rev.*, **67,** 317 (1967).
3. D. J. Pasto, S. K. Chou, A. Waterhouse, R. H. Shults, and G. F. Hennion, *J. Org. Chem.*, **43,** 1385 and 1389 (1978).
4. M. C. Caserio and R. E. Pratt, *Tetrahedron Lett.*, 91 (1967).
5. For a review on the effect of transition metal catalysis in Grignard reactions see E. Erdik, *Tetrahedron*, **40,** 641 (1984).

PROBLEM 31

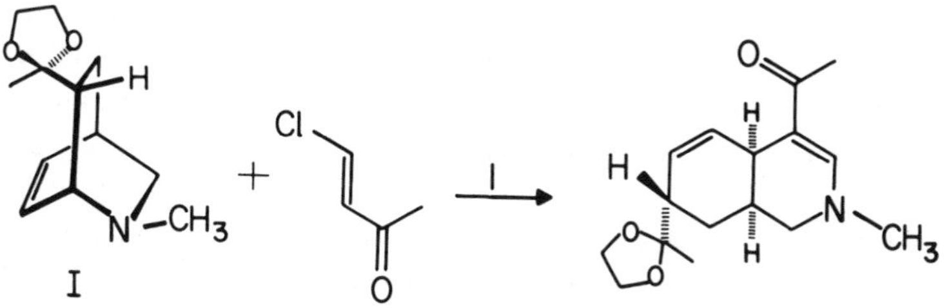

I + II $\xrightarrow{I}$ III (61%)

I = K_2CO_3, THF, 25°, 1.5 hr

---

1. P. S. Mariano, D. Dunaway-Mariano, and P. L. Huesmann, *J. Org. Chem.*, **44,** 124 (1979).

PROBLEM 31
Importance of Geometric Factors of Intermediates on the Stereochemical Outcome of a Reaction

Once again, the powerful technique of molecular fragmentation of the target compound, as indicated in Scheme 31.1, clearly indicates that, if fragment **A** is likely to come from β-chloroenone (**II**), then the other fragment **B** may be traced back to the bicyclic structure **I.**

From this scheme, various coupling routes of the two involved fragments may be proposed. Some additional background information may be helpful at this point. The reaction of β-chloroenones with amines has been known for some time to give β-amino-α,β-unsaturated ketones.[2] Therefore, the C–N bond formation of the enamino ketone portion of ring **B** of **II** would be accommodated to this precedent in which case the quaternary salt **IV** would result (see Scheme 31.2).

The formation of the second bond from which ring **B** results involves a more complex process, and a number of mechanisms may be advanced to account for it. First, a [2,2,2]bicyclic structure with a positive charge on the molecular skeleton is a strong stimulus for proposing a heterolytic bond separation of the C–N linkage to give **V,** a contention that is supported by the appearance of a relatively stable allylic carbocation. Resonance structure **VI** would place this positive center at an ideal position for its capture by the enamino ketone π system. The consolidation of the resulting C–C bond would not only yield compound **III,** but more importantly, in the proper cis-ring fusion.

There is, however, one shortcoming in this conception: The allylic cation

I

R= CH_3

B

III

II

A

SCHEME 31.1

SCHEME 31.2

in the six-membered ring places five atoms of this ring nearly on the same plane. In addition, the enone chain stretches out from this ring in a conceivably pseudo-equatorial configuration. This chain is long enough to bend itself and favor a second mode of π-orbital interaction, which is represented by structure **VIII,** which would lead to trans-fused compound **X** (see Scheme 31.3). This finds no correspondence with actual experimental fact. Examination of molecular models for both cyclizations does not warrant the stereoselectivity observed.

Second, a mechanism based on a chloride ion inducing the same initial ring disconnection via an S_N2' attack is not devoid of some consistency. The chloride ion is presumably tightly bound to the quaternary amine salt as a counter-ion, and is therefore available for reaction with entropic factors in its favor. Now, if this chloride ion were to approach **IV** by its exo side (route **A,** Scheme 31.4) the ensuing enamino ketone (**IX**) could then displace this chloride in S_N2 fashion from the underside and yield the observed cis-ring fusion. Unfortunately, this is not a reasonable proposition, because in this kind of bicyclic structure with a bulky ethylene ketal substituent just above the C=C bond involved, the attacking species usually abandon the exo approach in favor of the endo approach. In fact this very circumstance allows for the convenient positioning of the enone on the endo side of **I** during the first step of Scheme 31.2. Then, if the attack by chloride were to proceed from the endo position the ensuing cyclization would necessarily finish in a trans-ring fusion, as portrayed in **XII** (route **B**).

A third alternative may also be conceived, if for a moment one disregards all the nonparticipant atoms in compound **IV**. What is left is just an allyl–vinyl amine, a molecular combination reminiscent of what is necessary to perform a Claisen-type rearrangement (see Problem 19). Application of this rationale to intermediate **IV** would lead to cis structure **III** in a smooth, concerted fashion. This transformation is known in the literature as the amino-Claisen rearrange-

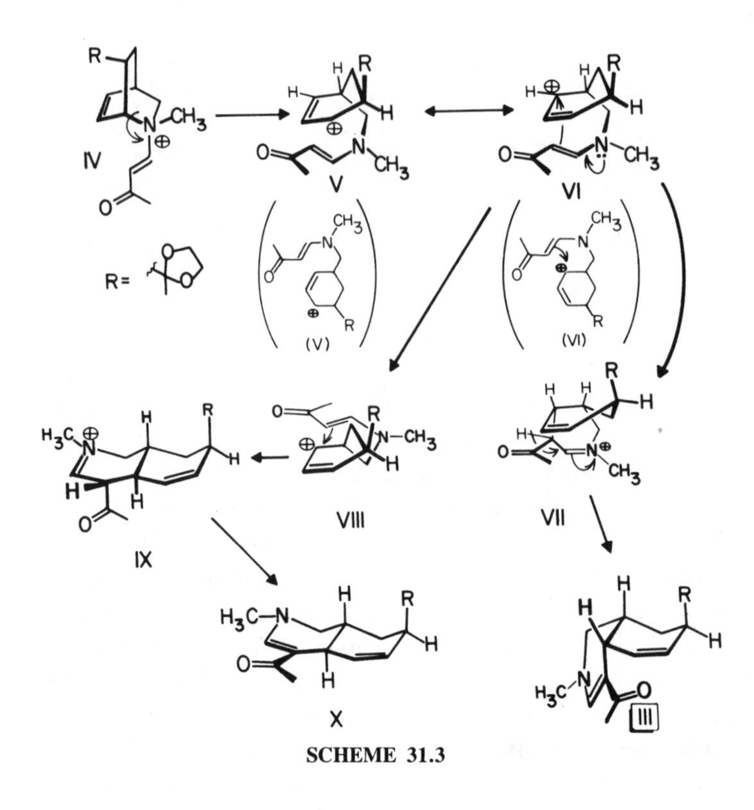

SCHEME 31.3

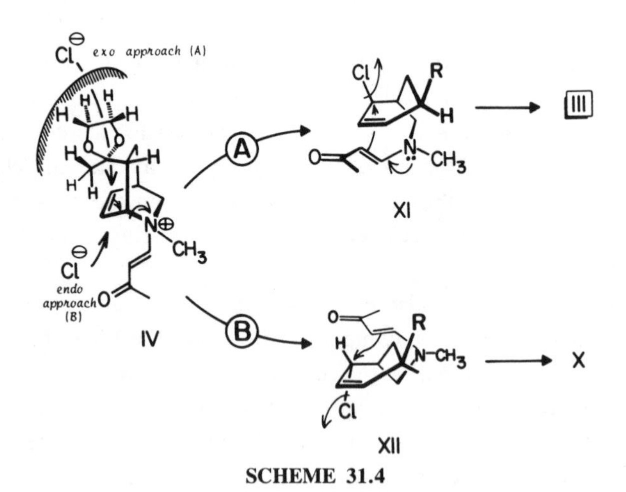

SCHEME 31.4

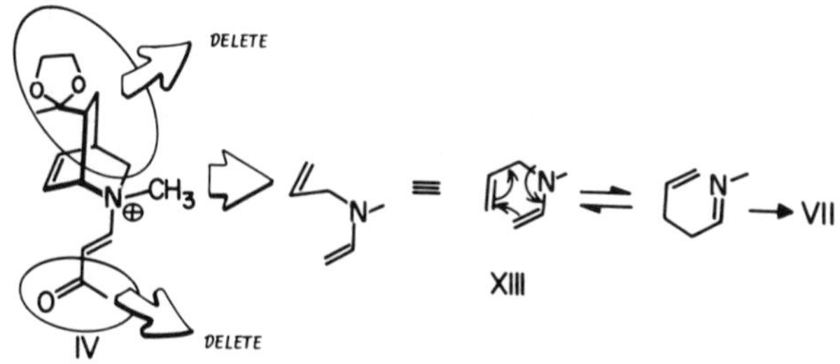

IV

XIII

SCHEME 31.5

ment, whose great potential appears at this date open to further research in mechanism and synthetic applications (see Scheme 31.5).[3]

Of the mechanisms discussed, the last option appears the closest to reality in view of the manner in which it accounts for experimental observations.

It will be instructive to comment briefly on the outcome of compound **XIV,** whose difference with respect to intermediate **IV** is the nonquaternization of the amine portion. Although the result of its exposure to **II** is the same in that tricyclic compound **XVIII** is produced, the amino-Claisen rearrangement cannot account for this process, since the protonation that would be expected to occur at nitrogen, thereby triggering the concerted transformation, surprisingly takes place instead on the oxygen atom of the carbonyl function (see Scheme

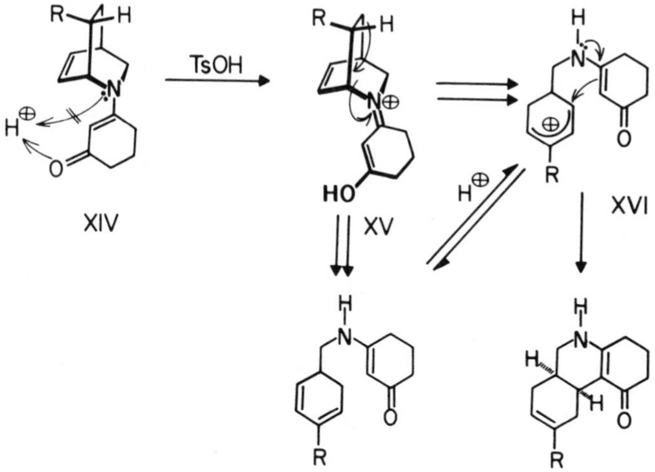

XIV XV XVI XVII XVIII

SCHEME 31.6

31.6).[4] Enol **XV** is certainly not prone to undergo rearrangement since the required vinyl amine would now be an α,β-unsaturated iminium ion. As a consequence, the route open to this intermediate (**XIV**) is only the β elimination of **XV** and the trapping of the resulting carbenium ion (**XVI**) with the vinylogous amide function.

REFERENCES

2. For a review see A. E. Pohland and W. R. Benson, *Chem. Rev.*, **66,** 161 (1966).
3. J. Corbier and P. Cresson, *C. R. Acad. Sci., Ser. C*, **270,** 2077 (1970) and references cited therein.
4. Simple enamino ketones typically are protonated only at the carbonyl group. See N. J. Leonard and J. A. Adamcik, *J. Am. Chem. Soc.*, **81,** 595 (1959).

PROBLEM 32

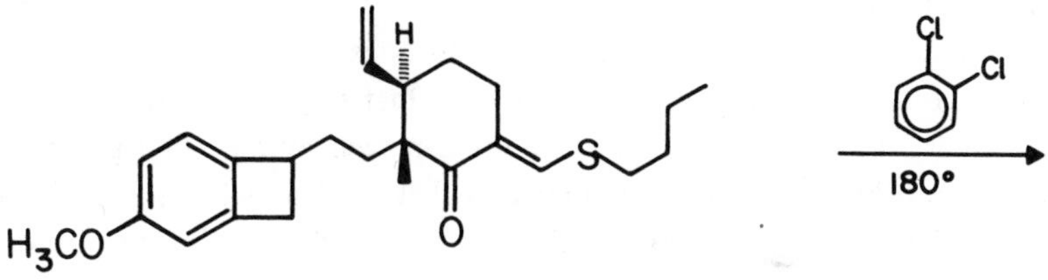

H_3CO

I

$\xrightarrow{\text{o-dichlorobenzene, } 180°}$

H_3CO

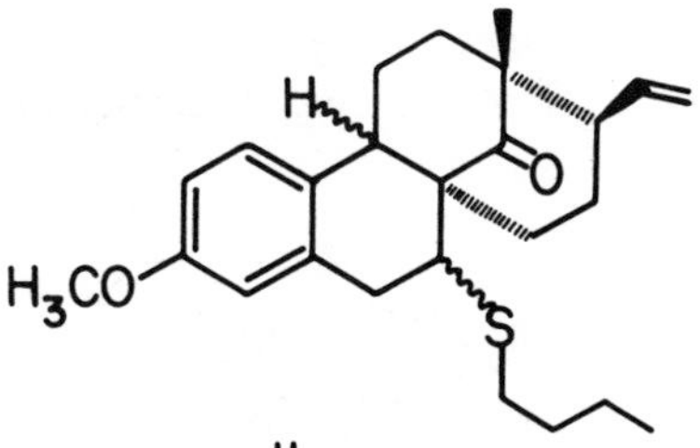

II

1. T. Kametani, H. Nemoto, H. Ishikawa, K. Shiroyama, and K. Fukumoto, *J. Am. Chem. Soc.*, **98,** 3378 (1976).

PROBLEM 32
Isolation of the Wrong Diels–Alder Adduct

Atom and bond budgets of starting material and product indicate that this transformation is simply an isomerization where two new C–C bonds have been created at the expense of one C=C bond. In the absence of an efficient leaving group, which is the case in **I**, this cannot be achieved. Therefore, the reaction mechanism must not be underestimated.

In spite of the fact that fragmentation analysis of **II** (Scheme 32.1) cannot foretell the nature of the sequence of events that separates starting and target compounds, it sheds some light on the role played by the vinyl group of the thiobutyl protecting fragment. It indicates that it is the stage from which the two new C–C bonds are constructed, while the rest of the 3-vinylcyclohexanone group remains intact.

It also becomes apparent from this analysis that the two tetragonal carbons of cyclobutene in **I** must have been activated at some point for them to become the complementary centers of C–C bond construction. This activation occurs by means of the cyclobutene ring disconnection. This fragmentation follows

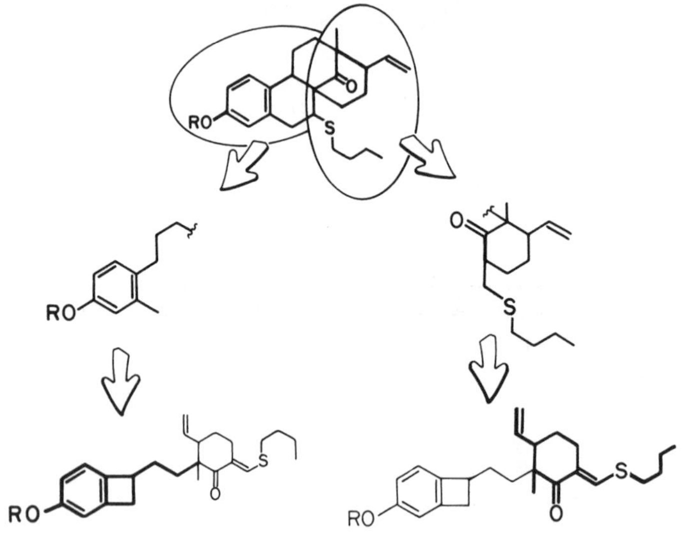

SCHEME 32.1

well established chemistry[2] and from it polyene enol ether **III** would result (see Scheme 32.2).

Intermediate **III** features the familiar combination diene–dienophile (the *n*-butylvinyl sulfide fragment). Consequently, it is prone to undergo a Diels–Alder cycloaddition in an intramolecular fashion.[3] As a consequence, tetracyclic structure **II** would be built in essentially one step.

Although the relative stereochemistry of the B,C ring fusion of compound **II** was not discussed in the original paper, it is always instructive to treat this point in cycloaddition reactions, and some valid predictions may be advanced on theoretical grounds only. Assuming that a transoid C=C bond is produced during the destruction of the cyclobutane fragment, the only model of inter-

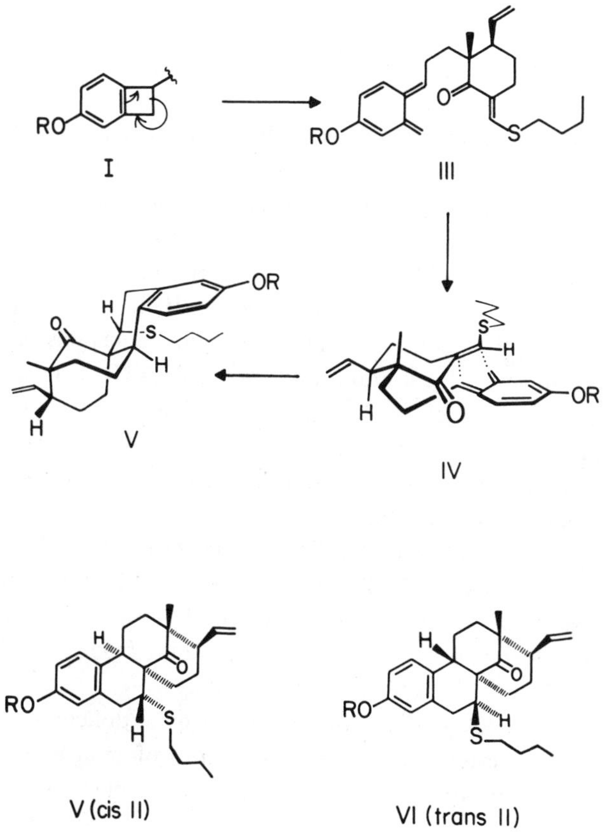

I III IV V

V (cis II) VI (trans II)

SCHEME 32.2

RO VII → RO VIII

RO IX → RO X

SCHEME 32.3

mediate **III** with the proper alignment of diene and dienophile that this writer was able to build or draw is the one represented in **IV.** Only the cis-ring fused structure **V** can be derived from it, thus making the formation of the trans derivative **VI** unlikely.

It is also of interest that the 3-vinyl substituent in the cyclohexanone ring might place itself in that position that allows for an effective interaction with the diene as well. This was in actual fact the original intentions of the authors in an effort to synthesize the useful estrone precursor **VIII** in a short sequence using an intramolecular Diels–Alder reaction (see Scheme 32.3). The projected scheme could be realized, however, by exclusion of the interferring thiovinyl group. The ensuing Diels–Alder cycloaddition led to tetracyclic compound **VIII** with the desired trans-B,C ring fusion, probably because a much better π-orbital overlap is achieved in conformation **VII** than in conformation **IX** that precedes the unobserved cis derivative **X.** The reason for stereocontrol may also be found in the participation of "the more stable exo transition state **VII** rather than the endo state **IX,** which has steric repulsion between the aromatic and the cyclohexanone rings."[1]

A molecular model of **IX,** unfortunately, is quite difficult—if not impossible—to construct for no other reason that, in spite of an artful drawing (ours here reproduces the original proposition closely), it cannot possibly achieve this precise conformation.

REFERENCES

2. I. L. Klundt, *Chem. Rev.*, **70,** 471 (1970).
3. For recent reviews on this active field of research see W. Oppolzer, *Angew. Chem. Int. Ed. Engl.*, **16,** 10 (1977); W. Oppolzer, "Intramolecular [4 + 2] and [3 + 2] Cycloadditions in Organic Synthesis," in *New Synthetic Methods*, vol. 6, 1979, p. 1, The Royal Society of Chemistry.

PROBLEM 33

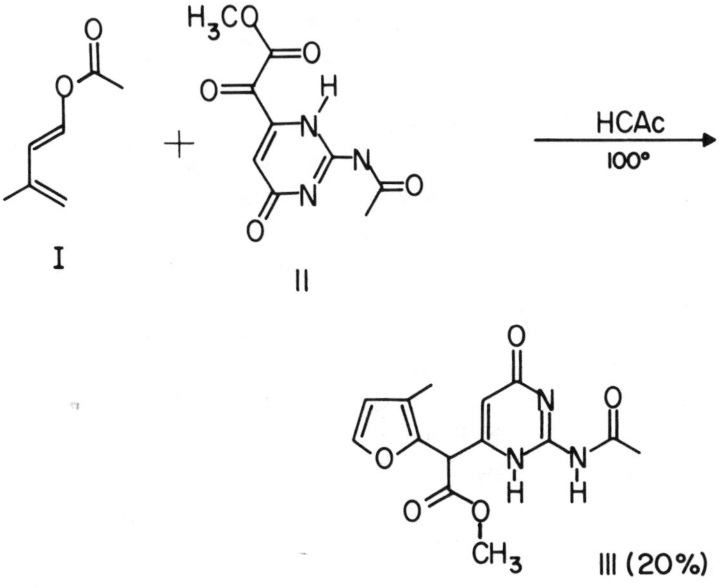

1. J. F. W. Keana and P. E. Eckler, *J. Org. Chem.*, **41,** 2850 (1976).

PROBLEM 33
An Unplanned Diels–Alder Cycloaddition

The classical Diels–Alder cycloaddition usually, though not always, requires electron donating substituents on the diene and electron withdrawers on the dienophile. This combination is present in the reacting compounds **I** and **II.** Were it not for the explicit representation of **III,** over 90% of the readers—I would guess—would argue that **I** and **II** ought to give the hydroquinazoline ring system **IV.**[2] This was also the original thought of the authors of this paper, who based their failed prediction in the smooth cycloaddition undergone by the closely related dienophile **VI** onto diene **I,** wherefrom compound **VII** was isolated (see Scheme 33.1).[3] The long run objective was to convert **IV** into the challenging and extremely poisonous tetrodotoxin **V,** a fish toxin.

Strong reasons would support this wrong assumption. On one hand, besides the only C=C bond of **II,** there appears to be nothing in this molecule reactive enough towards the diene, save perhaps for the α-keto ester function. However, most people believe that the only way to drive a carbonyl to react in a Diels–Alder fashion is by placing it in the diene portion such as in the synthesis of dihydropyran derivatives (see Scheme 33.2).

Although one such group is observed in **II,** in principle it would not find the complementary dienophile in compound **I** since the latter obviously plays the role of dienophile. However, that this is not always the case is exemplified

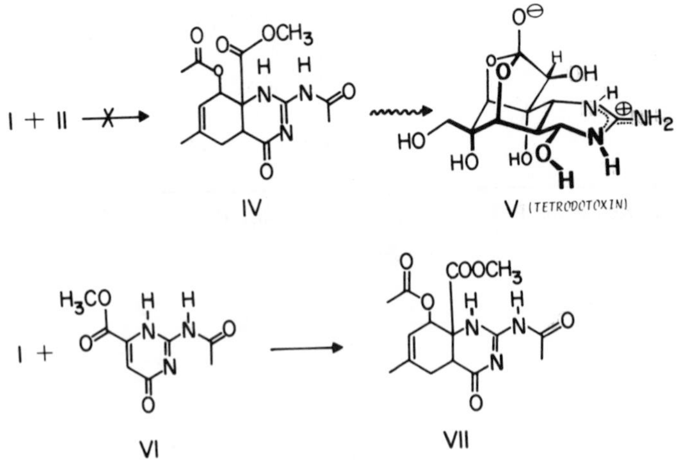

IV

V (TETRODOTOXIN)

VI

VII

SCHEME 33.1

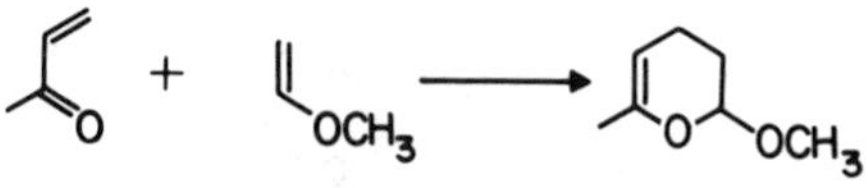

VIII IX

SCHEME 33.2

by the use of diethyl oxomalonate, mesoxalonitrile, and formaldehyde as dienophiles in Diels–Alder reactions with 1,3-dienes.[4] Despite that, none of these simple model compounds find an adequate analogy in compound **II.** Its α-keto ester function would be a vinylogous version of oxomalonate.

As a consequence, the way to a [4 + 2] cycloaddition would be open, with the linking point being this carbonyl group. If dihydropyran **X** were formed (note the position of the C=C bond as compared to **IX**) and no other reagents were to be added, then compound **X** should contain all the structural elements

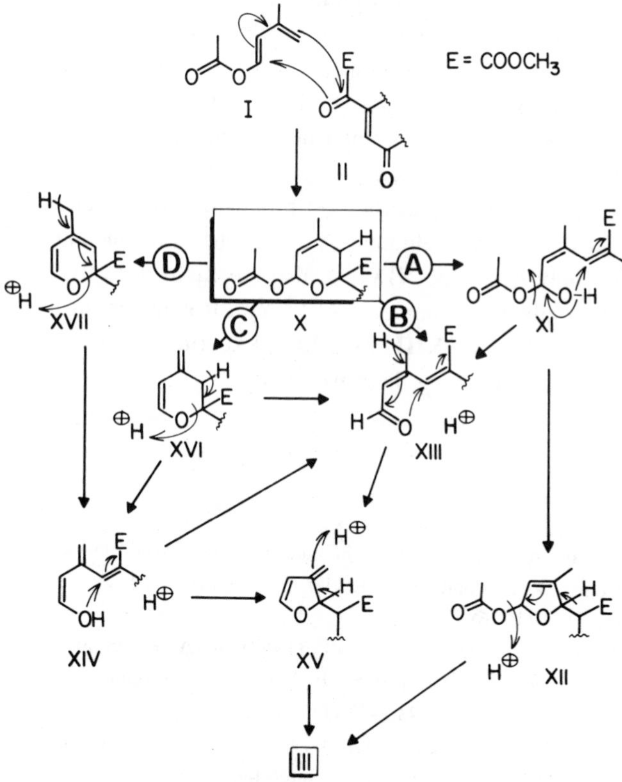

E = $COOCH_3$

SCHEME 33.3

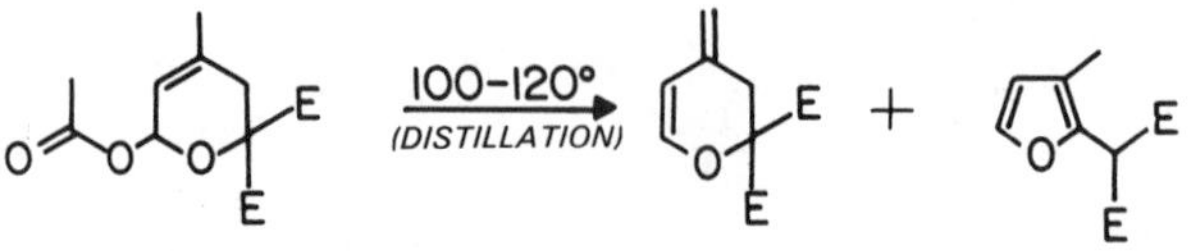

100–120° (DISTILLATION)

XVIII $E = COOCH_3$ XIX (50%) XX (?)

SCHEME 33.4

required for its ring contraction to the target furan nucleus. In fact, the pivotal intermediate **X** may do so in at least four different ways depending on whether the initial protonation that presumably unchains the process occurs at the dihydropyranyl oxygen (route **A** of Scheme 33.3) or at the oxygen atoms of acetate (routes **B, C,** and **D**) with concurrent 1,4 elimination of acetic acid (various combinations available). All these alternatives nevertheless converge towards the Michael-type addition of the α,β-unsaturated ester moiety that is derived from this, namely, **XIII** and **XIV** yielding **XV** while **XI** furnishes **XII.**

The participation of an aldehyde intermediate was suggested by the detection of a CHO proton in the nmr spectrum of a crude distillate in a related experiment.[1] This observation suggests that dienyl aldehyde **XIII** may be on the way to compound **III.** This should not be surprising since every reaction pathway of those just postulated, converge towards this aldehyde.

Route **C** stands probably as the best proposition if one takes into consideration the ease of acetic acid elimination undergone by compound **XVIII,** where we obtain not only the diene **XIX** but also the furan derivative **XX** to some undetermined extent (see Scheme 33.4).[1,5] Yet, it is uncertain if this intermediate evolves toward **III** via **XIII** or via **XIV.** Experimental data presently available cannot distinguish among these pathways.

REFERENCES

2. I did. As for the remaining 10% of the audience, they would not know what to do!
3. J. F. K. Keana, J. S. Bland, P. E. Eckler, V. Nelson, and J. Z. Gougoutas, *J. Org. Chem.*, **41,** 2124 (1976).
4. D. G. Kubler, *J. Org. Chem.*, **27,** 1435 (1962); R. A. Ruden and R. Bonjouklian, *J. Am. Chem. Soc.*, **97,** 6892 (1975); E. J. Corey, T. Ravindranathan, and S. Terashima, *J. Am. Chem. Soc.*, **93,** 4326 (1971).
5. See also E. N. Marvel, T. Chadwick, G. Caple, T. Gosink, and G. Zimer, *J. Org. Chem.*, **37,** 2992 (1972) and references cited therein.

PROBLEM 34

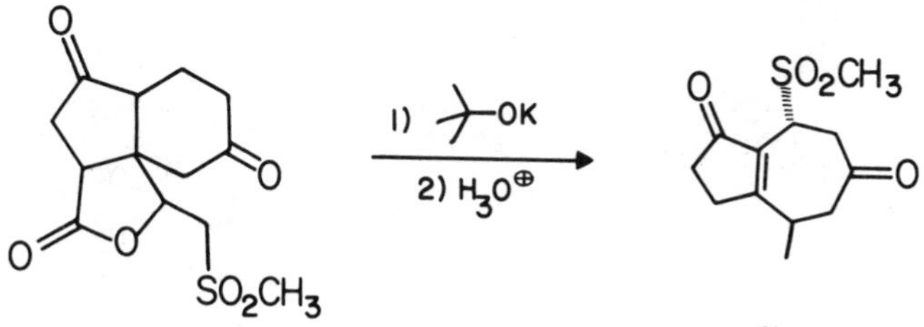

$$\text{I} \xrightarrow[\text{2) } H_3O^{\oplus}]{\text{1) } t\text{-BuOK}} \text{II}$$

I

II

---

1. J. Fayos, J. Clardy, L. J. Dolby, and T. Farnham, *J. Org. Chem.*, **42,** 1349 (1977).

PROBLEM 34
A Little of Everything: Various Carbanions and Enolates Yielding Ring Expansion

In contrast to what we have practiced in this writing, reasonable solutions to complex mechanistic situations can sometimes be found after just a few explanatory words. In this reaction, not an easy one to foresee to be sure, a probable mechanism unfolds after the recognition of very few facts.

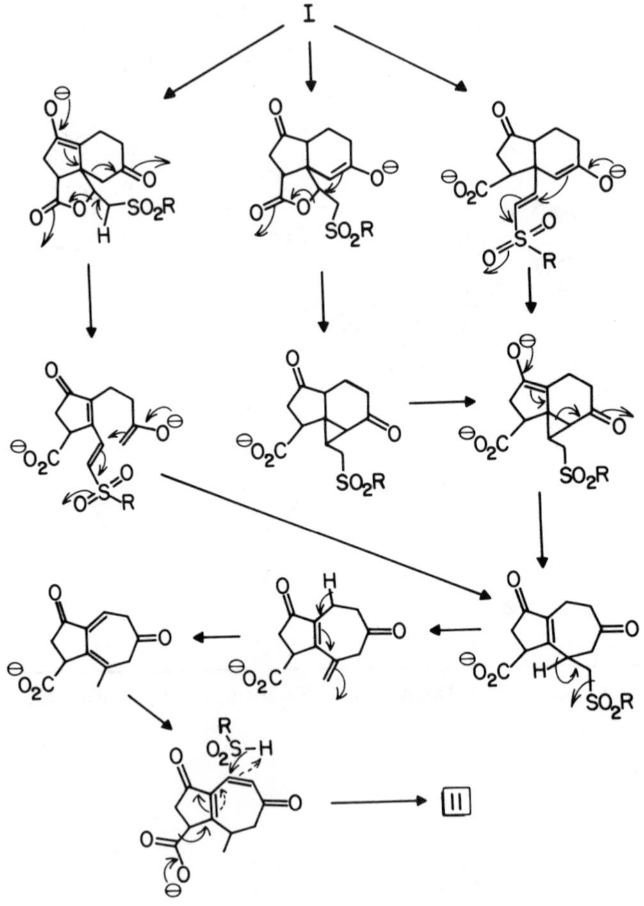

SCHEME 34.1

All that is taking place is ring expansion, lactone ring opening with decarboxylation, and finally a local oxidation of the carbon atom at which the sulfone substituent in **II** is placed.

For the ring expansion step an extra carbon is needed, of course, and the γ carbon of the lactone ring appears to be the best candidate. It not only bears a relatively labile C–O bond that upon disconnection might evolve into the carboxylate unit required for decarboxylation, but this carbon is also proximal to the α anion of the cyclohexanone portion of molecule **I** that is expected to appear in strong basic medium. The union of these two carbons will produce a cyclopropyl ketone whose central C–C bond might break up, prodded by the carbonyl unit, to yield a seven-membered ring system.

This in turn implies that the secondary methyl group in **II** is the same methylene carbon on which the methyl sulfone is substituted in **I.** If so, this methylene must be reduced at some point in the sequence, and might well be the reduced species needed for the local oxidation mentioned previously—no oxidizing or reducing reagents in the medium. As Scheme 34.1 shows, the double bond reorganizations and hydrogen shifts that automatically accompany the aforementioned steps achieve this goal directly. Our scheme contains other alternatives, which are drawn explicitly enough to make any further comment unnecessary.

PROBLEM 35

I ⟶ III (reagents: I)

I = II, piperidine, ROH, 0 - 20°

R = C_2H_5

1. W. D. Jones and W. L. Albrecht, *J. Org. Chem.*, **41,** 706 (1976).

PROBLEM 35
Acetoacetate Ester and the Knoevenagel Condensation

The combination of a moderate base—piperidine—and an active methylene compound such as **II** makes it rather obvious that something in the way of a Knoevenagel-type condensation may be taking place with the starting keto aldehyde **I**. This simplistic view, however, becomes more complicated, if the number of carbons of starting materials and product is compared. While the former include 16 carbon atoms, the latter contains 20. At this point one may think of 2 mol of β-keto ester **II** being used. In that case, nevertheless, there would be too many carbons, 22 in all. While decarboxylation is a convenient way of getting rid of excess carbons—and oxygens—some difficulty would be found in attempting to solve the question of the remaining two surplus carbons, because in this reaction the two carboxylate functions that would be introduced by 2 mol of **II** are present in the target product.

Fragmentation analysis of **III** (Scheme 35.1) portrays the distinct participation of the 2 mol of ethyl acetoacetate. The scheme also helps in concluding that the two ejected carbons are, in all probability, associated to an acyl group of only one molecule of **II.** Consequently, if the ketone carbon atom also departs, the mechanism must include an acidic type cleavage, the counterpart of decarboxylation that allows for the expeditious removal of acyl groupings.

As for the coupling of **I** and 2 mol of **II,** benzopyranone (**I**) features three possible sites amenable to nucleophilic attack: The two carbonyls and the tri-

SCHEME 35.1

gonal carbon β to these carbonyls, denoted with an asterisk in Scheme 35.2, owing to its double conjugation with two electron withdrawing functions. The question now becomes—at which carbon does **II** get coupled first? The experimental fact that, change of the base to sodium acetate and acetic anhydride causes the unique production of compound **IV** (62%) suggests that the addition reaction occurs at the aldehyde carbon. Other reports converge towards the same conclusion.[2] Besides, aldehydes are generally more electrophilic than ketones.

At this point the starred carbon would experience additional electron withdrawal since it would now be in direct conjugation with three carbonyls, once the elimination of water occurs. This condition will increase its chances to override competition from the other three electrophilic centers present in this intermediate (**IV**) for the second mole of acetoacetic ester anion. The components of the resulting crowded enol ether derivative **V** are favorably disposed

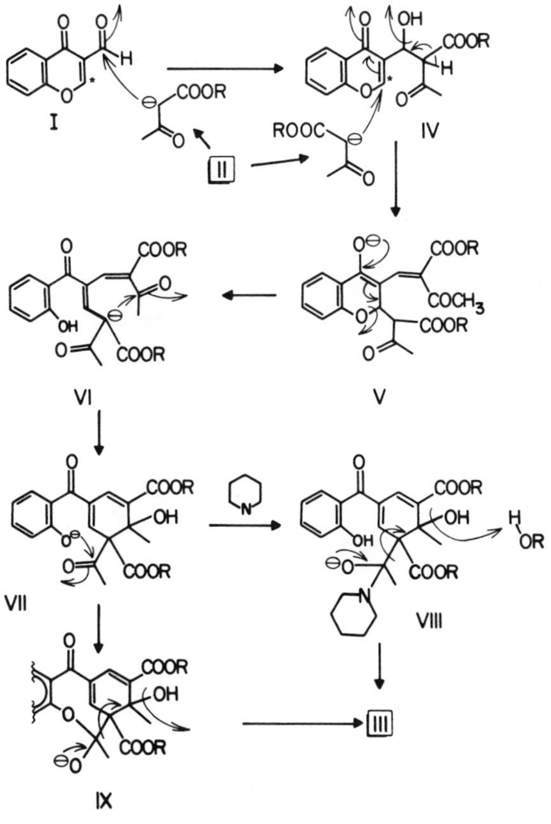

SCHEME 35.2

to promote a retro Michael-type disconnection of the pyranone ring. This operation would not only generate the desired phenol moiety, but in addition, considerable steric interaction would be mitigated. In such circumstances a kineticist would expect *steric acceleration.*[3]

The intramolecular condensation step expected for this species would lead to aldol derivative **VII** whose inability to eliminate the elements of water that would furnish the second aromatic ring leaves the way paved for the acidic cleavage mentioned earlier. For this cleavage to occur, an external nucleophile is required. In principle, piperidine, although it is a weak nucleophile, would be in a position to attack the ketone as depicted in **VIII** as it does during the production of enamines from ketones and aldehydes. However, the attacked carbonyl entity in **VII** is a tertiary ketone, which would be labile only with strong nucleophiles (see Scheme 35.2). This fact, and the advantages derived from its relative position, would favor the alternative attack on this ketone by the phenoxide ion. This would yield an eight-membered ring system **IX,** which would undergo fragmentation to the final product **III.**

REFERENCES

2. H. Harnish, *Justus Liebigs Ann. Chem.*, **765,** 8 (1972); see also A. Nohara, H. Kunki, T. Saijo, K. Ukawa, T. Murata, M. Kanno, and Y. Somo, *J. Med. Chem.*, **18,** 34 (1975) and references cited therein.
3. See, for example, I. Martin and G. Chuchani, *J. Phys. Chem.*, **85,** 3902 (1981).

PROBLEM 36

I → II (69%) + III (10%) + IV (15%) + V (trace)

I = $LiN(i\text{-}C_3H_7)_2$, THF, -78 to +20°

1. A. G. Schultz and M. H. Berger, *J. Org. Chem.*, **41,** 585 (1976).

PROBLEM 36
α Ester Anions and Their Several Ways to Self-Condensation

The beauty of this reaction lies in the fact that nearly all the facts needed to elucidate the mechanism are, in one way or another, in the products. Although the formation of **II** might seem somewhat tantalizing at first, a second glance will reveal that simply isomerization of **I** will suffice to account for it. A rather unusual isomerization, however, because activation of the α carbon of the ester as a nucleophile and introduction of formaldehyde (from where?) at this carbon need justification. The first argument may be reformulated as the formation of an ester enolate, which is made possible by the advent of lithium amide *superbases* such as lithium diisopropyl amide (LDA) in aprotic tetrahydrofuran (THF)–hexamethyl-phosphoramide (HMPA) solvent mixtures.[2] The participation of an ester enolate is emphasized by the formation of condensed diester **IV**.

Furthermore, by assuming that the methoxy unit maintains its integrity through the reaction, the formaldehyde fragment mentioned earlier must proceed from the oxymethylene unit of **I.** At this stage it is timely to focus one's attention on product **III.** It is a dimeric form of dimethyl ketene. This ketene must come from the ester enolate of **I** (see Problem 21). Therefore, the simultaneous generation of dimethyl ketene **VII** and formaldehyde is conceivable.

Up to this point nothing is really novel in this reaction.[3] Ketene generation is in fact one of the early recognized competing processes that affect acetoacetic ester condensations. This reaction is customarily ignored, however, since it consumes only a few percent of the keto ester, owing to the greater difficulty of the departure of alkoxide ion relative to the fragmentation of **I.** In addition, it reverses back to the starting ester upon attack by alkoxide.

If a plain and simple ketene derived from a β-keto ester anion can efficiently add methoxide ion to give back the ester anion, what would prevent dimethyl

VIII → IX → X

SCHEME 36.1

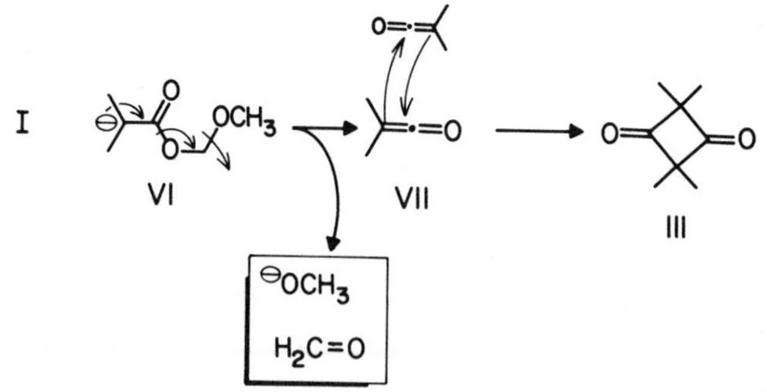

SCHEME 36.2

ketene **VII** from doing so? Besides the molecular diffusion that would reduce the local concentration of methoxy, no other nucleophile, least of all the bulky lithium base, would be available to compete for this ketene. Furthermore, the incorporation of methoxy would result in a new ester enolate **XI** whose addition to formaldehyde would yield **II** (see Scheme 36.3).

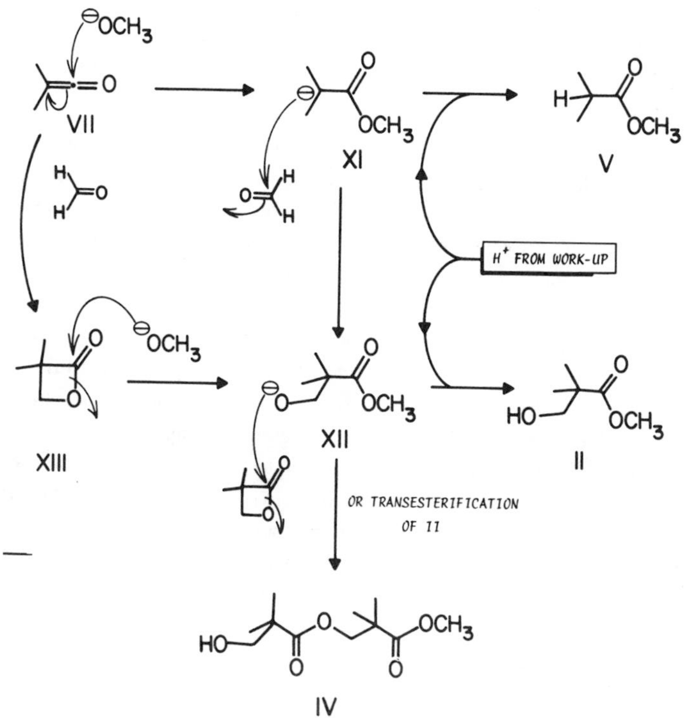

SCHEME 36.3

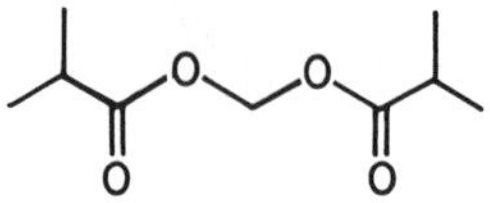

XIV

Along the same lines, competitive proton transfer (from where?)[4] would lead to **V** while direct transesterification would furnish compound **IV.**

Conceivably, the addition of formaldehyde and methoxide ion in reverse order, namely, by way of the construction of β-lactone **XIII,** would also be imaginable. Its justification would bear some difficulties, however, not only because of the intrinsic instability of **XIII** that would make this route less competitive, but also because of the appearance of methyl ester **V** that would militate against it. In addition, further evidence that the attack of methoxy precedes the intervention of formaldehyde is provided by the production of mostly isobutyric acid from treatment of **XIV** with LDA.[1]

REFERENCES

2. M. W. Rathke and A. Lindert, *J. Am. Chem. Soc.*, **93,** 2318 (1971).
3. The novel and instructive feature of this reaction is the excellent control exerted over this formerly undesirable reaction, which is achieved by careful manipulation of reaction conditions.[2] It is this attitude that has produced many useful syntheses out of calamitous and failed experiments, something rewarding enough to invite one to meditate over the convenience of a chemical *Journal of Wrecked Results* or, better, *Acta Chemica Kaputnika!*
4. In a nonprotic solvent containing very powerful bases such as LDA, it is hard to find a protic source capable of trapping the free anion **XI,** in particular if this species is not in equilibrium with the neutral ester **I,** since the conversion of esters to the corresponding enolates in LDA–THF–HMPA is essentially quantitative.[2] Moreover, it has been shown repeatedly that quenching enolates of simple esters with deuterium oxide yields a 1 : 1 mixture of protic and deuteric derivatives, even under rigorous exclusion of moisture.[5] Are the protons from glass walls or from the solar wind?
5. See, among other papers, M. E. Alonso, H. Aragona, A. W. Chitty, R. Compagnone, and G. Martin, *J. Org. Chem.*, **43,** 4491 (1978).

PROBLEM 37

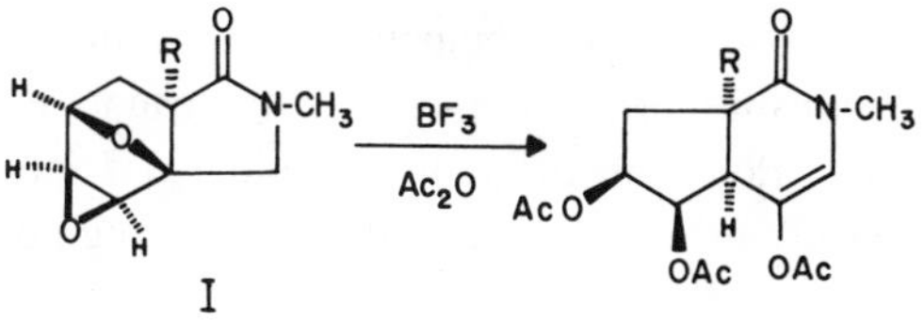

I

$R = C_6H_5$

II

1. H. W. Gschwend, M. L. Hillman, B. Kisis, and R. K. Rodebaugh, *J. Org. Chem.*, **41**, 104 (1976).

PROBLEM 37
Simultaneous Contraction and Expansion of a Carbocycle

Before analyzing this problem it will be instructive to review the interesting story of compound **I.** This molecule is just the *m*-chloroperbenzoic acid oxidation product of olefin **III** (see Scheme 37.1). This amide in turn was constructed by means of an intramolecular Diels–Alder cycloaddition as an alternative entry into the isoindoline derivative **VI.** Although its synthesis would be nothing more complex than a repetition of the reaction of **IV** yielding **III,** that is, **V** leading to **VI,** it simply did not work in the temperature range −60–230°C. Interestingly, it was shown that when the synthesis of **VI** was attempted by an independent route, namely, lithium aluminum hydride reduction of **III,** not a trace of **VI** survived, furan **V** being the only isolable material. As a consequence, one might conclude that compound **VI** is an unusual example of an apparently unobjectionable, orthodox molecule that cannot exist between −60 and 230°C. In connection with the transformation of **I** into **II,** its apparent complexity vanishes after the recognition of the following points:

1. the molecular backbones of **I** and **II** both have the same number of carbon atoms. Therefore, the change of ring sizes must be a ring expansion occurring at the expense of ring contraction of the complementary portion.

R O N-CH$_3$ I
R O N-CH$_3$ III
O R N-CH$_3$ IV
R N-CH$_3$ V
R N-CH$_3$ VI

SCHEME 37.1

2. the carbon skeleton of **II** has one more oxygen functionality (one of the three acetoxy units) than that of **I.** This calls for an oxidative process in the absence of oxidative reagents. The alternative is oxidative C–C bond breakage of the carbon nucleus. In turn, this requires a strong driving force such as extreme bond polarization in the neutral starting molecule (presently absent), or a carbenium ion directly derived from it.

Favorable conditions for the formation of carbenium ions include strong Lewis acid media, in addition to Lewis bases in the substrate. Boron trifluoride provides the first condition. As for the second, the Lewis base linking site of **I** is not one but two: The oxygen atom bridges. Of these two, the epoxide is probably the more favored since, on one hand, the more pronounced *s* character of the oxygen in a three-membered ring, as compared to the five-membered ring, would concentrate more electron density on the epoxide, thus making it more basic. On the other, the association of this oxygen atom with boron trifluoride would trigger ring opening and the formation of the carbenium ion that is likely to assist the C–C bond disconnection mentioned earlier. The three-dimensional representation of **I** (see Scheme 37.2) allows one to see that this

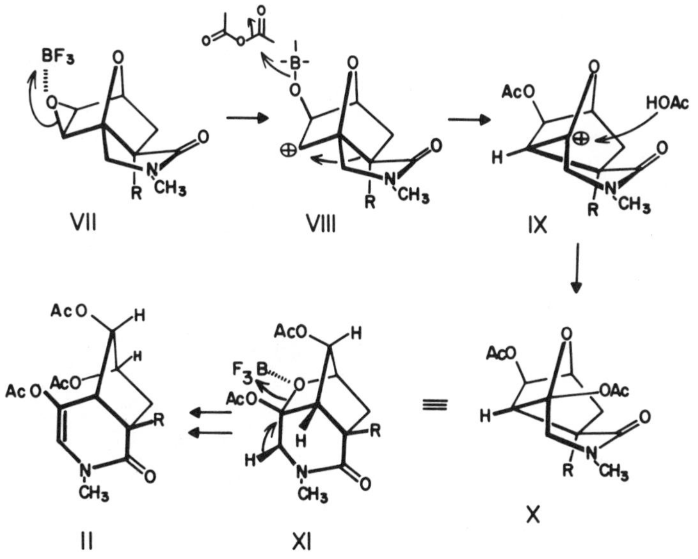

SCHEME 37.2

cation (**VIII**) lies in a preferred position for inducing the deep-seated rearrangement that is typical of [2.2.1]heptane ring systems. This transformation would automatically bring about ring expansion of the lactam unit and the simultaneous contraction of the carbocycle of **I.** Then, the resulting carbenium ion **IX** would be attacked by acetate from the medium. Proton elimination with simultaneous boron trifluoride induced collapse of the remaining oxygen bridge would finish the sequence.

PROBLEM 38

H_3CO I + ROOC-C≡C-COOR II ($R = CH_3$) → Δ, 65 hr (LOW YIELD)

III (50%) + IV (50%)

---

1. S. S. Hall and A. Duggan, *J. Org. Chem.*, **39**, 3432 (1974).

PROBLEM 38
A Carbenium Ion Promoting the Destruction of a Ring to Make a New One

The chemistry of 2-alkoxydihydro-3-*H*-pyrans such as the starting material of this problem is usually plagued by uncontrolled polymerizations, almost inevitable hydrolysis in the presence of traces of moisture, and so on. However, no matter how difficult these compounds are to handle, they are intriguing enough to deserve close attention.

The present reaction, which appears in the second of a series of provocative papers,[2] is an example of these difficulties, not only because of the low yield of recovered products but also because of the operation of an unexpected course of reaction. This is represented by product **IV.** This compound shares with starting material **I** the six-membered carbocycle, the β-keto ester fragment, and a C=C bond, although the latter is in a different position than that occupied by the acetylene group in **II** (between two carbomethoxy groups).

A first hint comes from the fact that the carbonyl carbon of the acyl substituent in **IV** may be the α carbon of the enol ether moiety in **I.** In fact, the vinyl ether would be converted into a methyl ketone by breakage of the bond between the acetal carbon and the endocyclic oxygen. In that case, the resulting oxycarbenium ion would serve as a linking position for the carbocyclization that eventually affords the cyclohexyl ring of **IV.**

It is also important to realize that the construction of the β-keto ester function in **VI,** whose predictable role as a stabilized carbanion can be exploited to build the carbocycle, requires the consolidation of a C–C bond between C-6 of compound **I** and a methyl acetate unit from **II.** This task cannot be accomplished without a preliminary step, since the carbons involved are hardly active in the nucleophilic sense at the outset. This prior step consists of the coupling of **I** and **II** as the consequence of the nucleophilic character of the enol ether fragment of **I** and the strongly electrophilic nature of the triple bond of **II.** The result would be zwitterion **V** (see Scheme 38.1).

While cumulene **V** might evolve directly into dimethylfumarate derivative **III** by proton transfer, alternatively, it might evolve into cyclobutene **VI,** whereby the desired C-6 methyl acetate unit is expediently formed. Such a structure has very little chance to survive in a hot reaction medium and three decomposition reactions are likely to take place: (1) return to starting materials; (2) a metathesis-type retro[2+2]cycloaddition to diene **VII,** something that finds

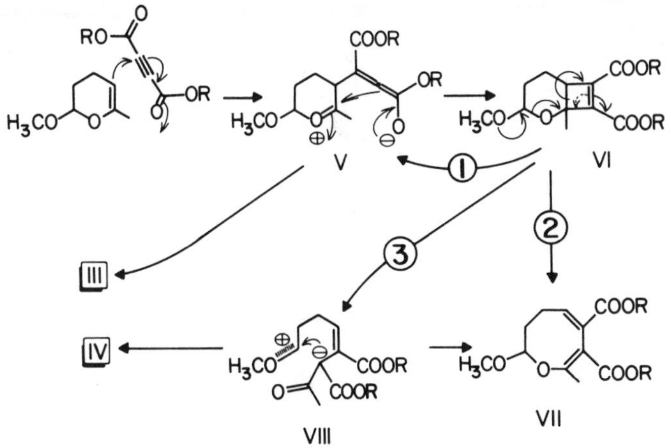

SCHEME 38.1

parallel in enamine chemistry; and (3) ring opening to intermediate **VIII** that would precede the target compound **IV.**

The latter possibility obeys the peculiar synergistic effect of the C-2 alkoxy substituent of **I** that makes these dihydropyrans unique. Without it, nothing of this occurs. These substrates have been used for the detection of positive charge development at C-6, among other things,[2] in additions of alkyl hypohalites and some diazocarbonyl compounds to the double bond,[3] and in the interesting fragmentation of norcarane derivatives illustrated in Scheme 38.2.[4]

It is also noteworthy that this alkoxy substituent is chiefly in the axial conformation in which its orbital contribution to the C–O bond disconnection is minimal. It has been suggested that a conformational change that places this methoxy group in the equatorial position occurs prior to ring scission.[2]

The central intermediate **V** would also be amenable to undergo a previous

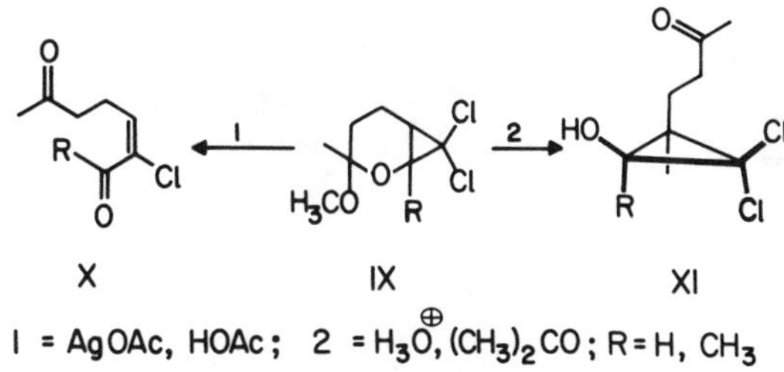

1 = AgOAc, HOAc; 2 = $H_3\overset{\oplus}{O}$, $(CH_3)_2CO$; R = H, CH_3

SCHEME 38.2

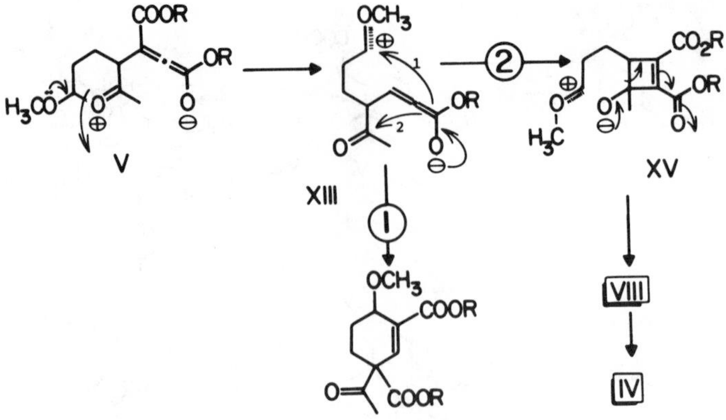

XIV
SCHEME 38.3

ring opening to intermediate **XIII,** and then continue towards **VIII,** hence **IV** via oxycyclobutene **XV** (see Scheme 38.3). However, the actual existence of **XIII** is shadowed by the absence of keto ester **XIV,** isomeric with **IV,** which would result presumably from a more favored cyclization of the ester enolate fragment onto the oxycarbenium ion than its intramolecular aldol condensation[4] with the acyl group to yield **XV.** Unfortunately, the experimental evidence required to make an intelligent choice among these various pathways is not available.

REFERENCES

2. The complementary articles of reference 1 are S. S. Hall and H. C. Chernoff, *Chem. Ind. (London)*, 896 (1970); A. J. Duggan and S. S. Hall, *J. Org. Chem.*, **40,** 2235 (1975); **42,** 1057 (1977); S. S. Hall, G. F. Weber, and A. J. Duggan, *J. Org. Chem.*, **43,** 667 (1978); G. F. Weber and S. S. Hall, *J. Org. Chem.*, **44,** 364, 447 (1979); J. H. Chan and S. S. Hall, *J. Org. Chem.*, **49,** 195 (1984).
3. Substrate **I** has been used recently to show for the first time the participation of charged species in certain additions of diazocarbonyl compounds to polarizable olefins. See M. E. Alonso and M. C. Garcia, *J. Org. Chem.*, **50,** 988 (1985).
4. For those who dislike an intramolecular aldol condensation yielding cyclobutene, see Problem 39 and E. D. Bergmann, D. Ginsburg, and R. Pappo, *Org. React.*, **10,** 191 (1959); A. Michael and J. Ross, *J. Am. Chem. Soc.*, **52,** 4598 (1930); N. E. Holden and A. Lapworth, *J. Chem. Soc.*, 2368 (1931).

PROBLEM 39

$$\text{I} + \text{NC-CH}_2\text{-C(=O)OCH}_3 \ (\text{II}) \xrightarrow[\text{HOCH}_3]{\text{NaOCH}_3} \text{III}$$

I   II   III

1. R. K. Hill and N. D. Ledford, *J. Am. Chem. Soc.*, **97,** 666 (1975).

PROBLEM 39
A Rigorous Michael Addition That Yields the Labeled Carbon at an Unexpected Position

In mechanistic reasoning, the most carefully conceived explanation may turn out to be a gross distortion of reality in spite of our perusal of supporting reference material and the rigorous application of Aristotelian logic. The reaction described in this problem is an example of such a situation.

Complications first arise when one notices the separation of nitrile and ester functions by a three-carbon chain in the product while there is only a methylene in this space in starting material **II.** This can only be understood in terms of a fragmentation–recombination process. That is, either the C–CN or C–COOMe fragment must maintain its integrity through the reaction, but not both.

The substitution of an acetic ester residue on the β carbon of cyclohexanone **III** points to a Michael-type 1,4 addition of the anion derived from **II** and base. This logical first step allows us to reformulate the problem in terms of a nitrile jump to the α carbon of **IV** in what is in essence a 1,3-nitrile shift. There are at least two ways to do this.

The first and shortest route (path **A** of Scheme 39.1) has the nitrile group as receptor of nucleophilic attack by a base-formed enolate. The resulting unstable intermediate **V** would then undergo a chemically sound fragmentation that would restore the nitrile function in the right place, directly yielding product **III.** Conversely, attack of the same enolate onto the ester (path **B**) would afford a relatively more stable uncharged cyclotubanone **VI.** This ketone features the components classically required for an acidic-type cleavage that is a characteristic shared by **VII** as well. Double fragmentation and proton transfer would lead to **IX,** wherefrom **III** would quickly result. Route **D,** although much less likely, would also explain the formation of **III** within the boundaries of the energetics of this process.

That the sequence **B,C** may be the closest representation of reality (with the evidence so far presented) is strongly supported by the formal mechanistic study based on ^{13}C and ^{18}O tracer analyses of the production of malonic ester derivative **XIII** from esters **XI** and **XII,** in which the cyclobutanone appeared as a necessary intermediate (see Scheme 39.2).[2]

However, in spite of obvious similarities the data of Scheme 39.2 cannot solve the **A** versus **B** dilemma. Labels should help without doubt, since the C–COOMe fragment of **II** remains intact in path **A** whereas alternative **B** in-

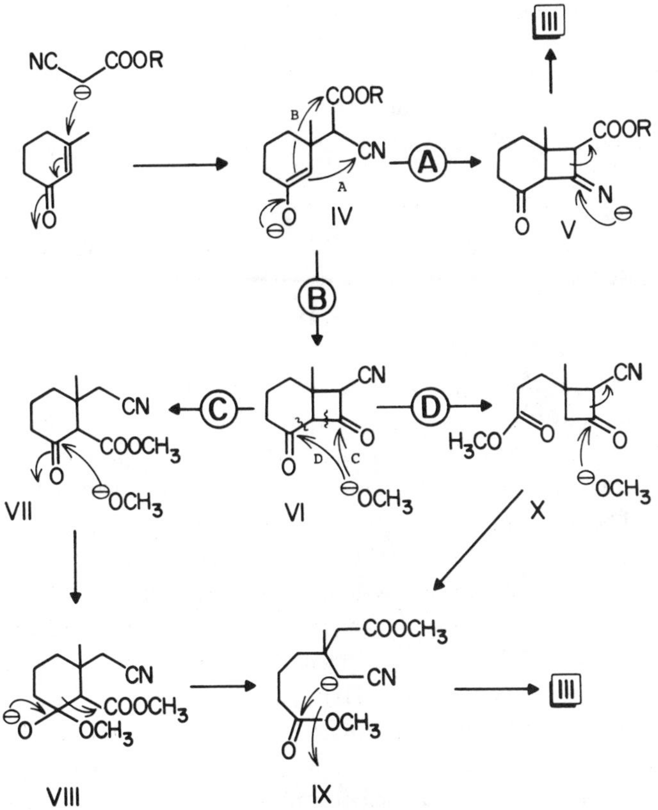

SCHEME 39.1

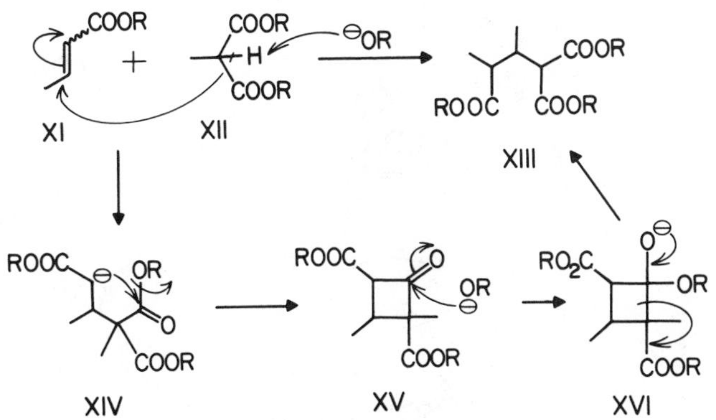

SCHEME 39.2

volves the conservation of the C–CN grouping. In fact, this desirable labeling experiment has been done,[1] although the tagged carbon was not the methylene carbon in cyanoacetate but rather the ester carbonyl carbon. Where would the reader expect this carbon in product **III?** After a careful structural analysis this starred carbon was found as part of the cyclohexanone carbonyl (see Scheme 39.3).

If we follow Scheme 39.1 again with this labeled carbon, we will soon discover it in the ester carbonyl of **III,** not in the ketone! The brainchild so adroitly delivered dies abruptly and our well woven supporting logic turns into mere rubbish.[3] Even alternatives **A** and **D** are wrong.

Neither **V** nor **VI** can lead to a cyclohexanone with the tagged carbon in the right position. Consequently, the correct mechanism must be a divergent route that stems from a previous intermediate. This species ought to be **IV** since up to that point it is the only one we have been able to conceive.

Only a few possibilities of evolution are open to this intermediate. One of these would be its equilibration with the other possible enolate **XVII** by way of proton transfer from the solvent followed by cyclization on the ester (route **E**) to give the bicyclo[2.2.2]octane system **XVIII** (see Scheme 39.4). One of the two keto functions would keep the labeled carbon if the other undergoes attack by the alkoxy anion followed by fragmentation as indicated. The end result of this would be the right compound—**III.**

There is still another pathway (route **F**) that accounts for the observed label. This proposition makes use of **XIX** as the key intermediate.

The only powerful argument against route **F** would be if compound **III** is unaffected by the action of base. Since it is the recovered product, putative compound **XX** should be stable as well, and therefore should be present in the reaction mixture. This is in opposition to actual experimental results.[1,4] A still

I + NC–CH₂–C*(=O)–OCH₃ → (A,B,C) ; (ACTUAL FACT)

SCHEME 39.3

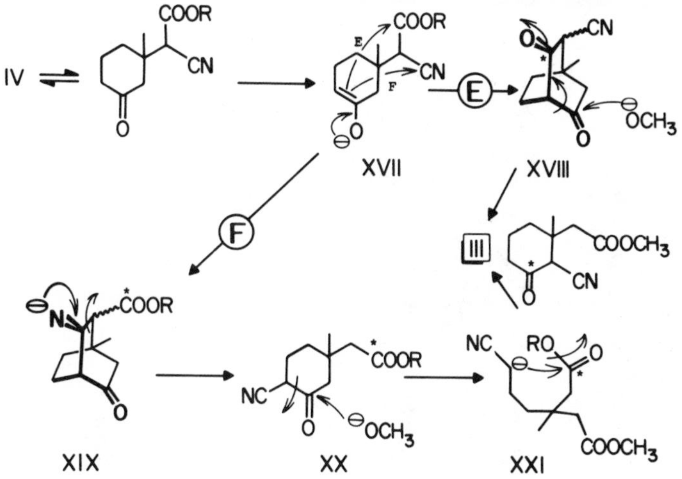

SCHEME 39.4

unrecorded experiment in which independently synthesized **XX** would be exposed to sodium methoxide in hot methanol may provide the necessary information.

REFERENCES

2. O. Simmamura, N. Inamoto, and T. Suehiro, *Bull. Chem. Soc. Jpn.*, **27,** 221 (1954); D. Samuel and D. Ginsburg, *J. Chem. Soc.*, 1288 (1955).
3. The **B,C** pathway was first proposed by Professor Johnson as a remarkable exercise in speculation. The confirming experiments were performed some 20 years later.[1] See P. R. Shafer, W. E. Loeb, and W. S. Johnson, *J. Am. Chem. Soc.*, **75,** 5963, (1953).
4. For a discussion about abnormal Michael additions catalyzed by base see E. D. Bergmann, D. Ginsburg, and R. Pappo, *Org. React.*, **10,** 191 (1959).

PROBLEM 40

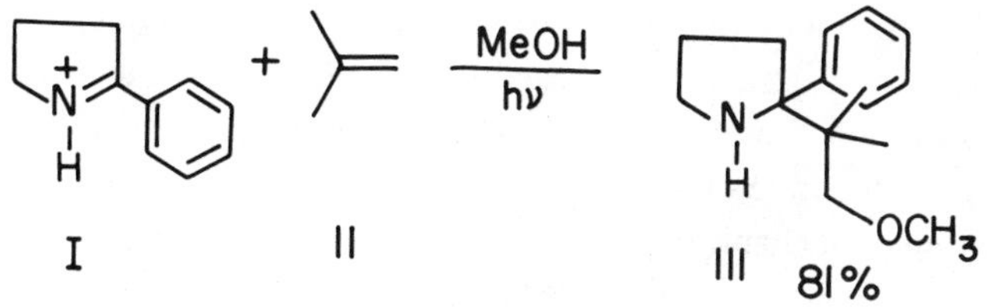

MeOH

$h\nu$

I + II → III 81%

1. J. L. Stavinoha and P. S. Mariano, *J. Am. Chem. Soc.*, **103,** 3136 (1981).

PROBLEM 40
One-Electron Transfer Process or Time for Controlled Confusion

This is an apparently simple reaction more appropriate for a freshman organic chemistry course than for the sophisticated graduate student. It may take several hours of hard labor, however, to find the actual solution if one uses one's chemical knowledge properly.

It may be clear that this reaction involves the formation of one C–C and one C–O bond during the linking of three molecules: **I, II,** and methanol from the solvent. The combination of an expectedly strong electrophile such as the pyrrolinium ion (**I**) and a very mild nucleophile represented here by isopropylene, falls within the familiar framework of a Prins -type reaction.[2] That process has a protonated form of a carbonyl derivative as the electrophile, usually an aldehyde of which the iminium ion would be a nitrogen homolog. However, in contrast with the classical Prins reaction whereby 1,3-dioxanes result by the inclusion of a second mole of the electrophile, in the present reaction 1 mol of methanol is incorporated instead.

One simple mechanism that would readily account for product **III** would follow an ionic route. A photoexcited form of **I** would be attacked by the olefin to form an intermediate carbenium ion that subsequently would be trapped by the methanol solvent. However, **II** is an unsymmetrical alkene and the ionic mechanism would favor the linking of **I** and **II** in a direction opposite to that required by the construction of **III.** That is, compounds **IV** and **V** should be the main products.

Nevertheless, the reaction is reportedly regiospecific, even with alkenes whose polarization is more strongly governed by conjugation with carbonyls than with mere alkyl substitution. For example, Eq. (1) illustrates the reaction of **I** with β-dimethylacrylate ester.[1] Consequently the ionic mechanism can be dismissed.

A second possibility that would be consistent with the regiochemistry ob-

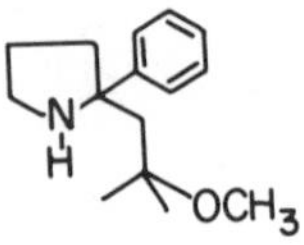

N H OCH$_3$

IV

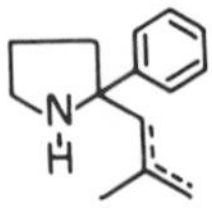

N H

V

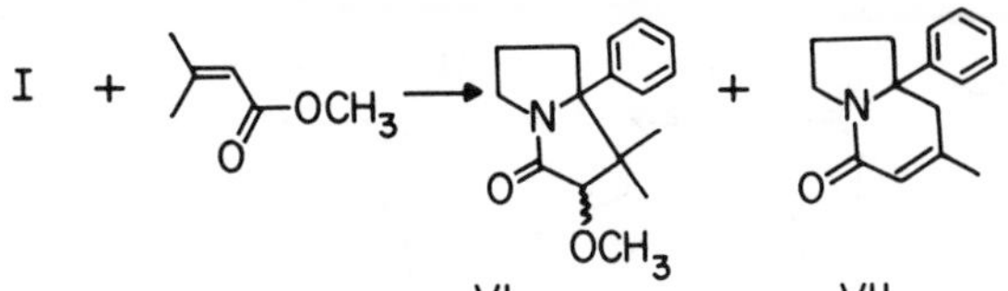

(eq. 1)

served would follow a free radical route. This sequence starts with a photolytically induced homolytic breakage of methanol to form methoxyl radicals. These radicals would propagate through the reaction system by way of radical attack on the olefin (alkenes make efficient radical scavengers) to form a moderately stable tertiary radical **VIII.** This species then would undergo radical coupling with **I** in the observed direction.[3]

A second piece of experimental evidence to be considered at this point is the photoaddition of **I** to a substrate that is sensitive to the appearance of a free radical in its vicinity, that is, the cyclopropyl ethylene derivative **X.** This sensitivity is expressed by its fast fragmentation to a pentenyl derivative **XIII** that would give rise to coupling compound **XIV.**

Only compound **XII** was observed,[1] which illustrates once more the anti-Markovnikov regioselectivity of the addition reaction. However, the divergent fate of intermediate **XI** must be governed by the relative rates of the two competing processes **A** and **B** (Scheme 40.2), so the radical mechanism cannot be dismissed on the basis of this experiment alone [Eq. (2) of Scheme 40.2]. It has been estimated[1] that the rate of step **B** is of the order of $k = 3 \times 10^5\ s^{-1}$ and the activation energy is only 13 kcal/mol. This means that if the mean life time of **XI**—before it is trapped by **I**—is shorter than 300 thousandths of a second, pathway **B** will be blocked and only compound **XII** will be produced.

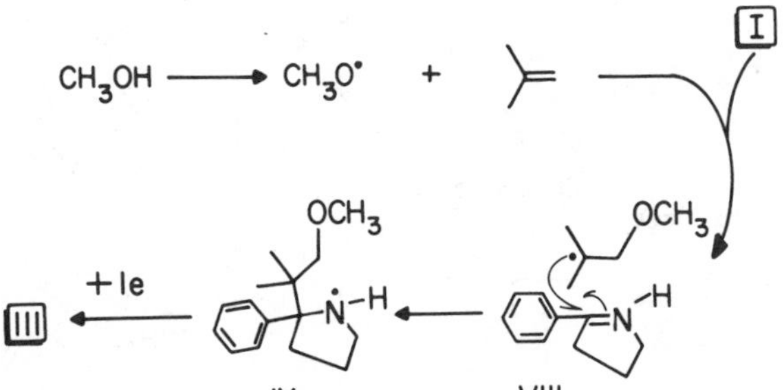

SCHEME 40.1

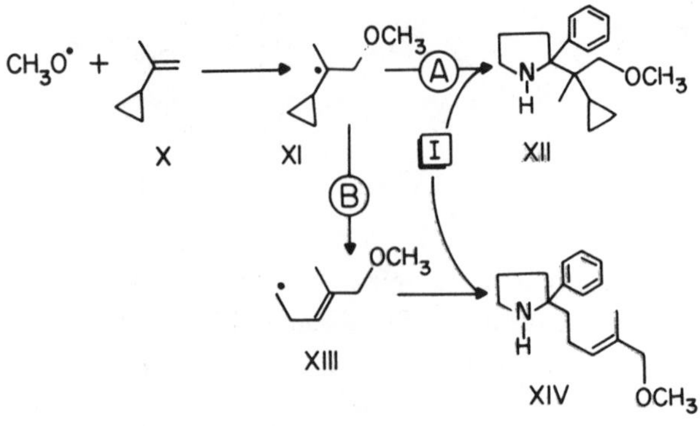

(eq. 2)

SCHEME 40.2

Therefore, the radical route seems to be a good alternative since the rate of reaction of **XI** with **I** might well be diffusion controlled, that is, it is very fast.

Information extracted from a meticulous library search, however, invalidates the radical mechanism.[4] In a recent short communication researchers report that the quenching rate of the fluorescence of compound **I** with electron-rich alkenes is extremely fast, close to the diffusion control limit. Conversely, fluorescence quenching by methanol is 100 times slower. In the present case, this means that photoexcited **I** would be expected to react much faster with isopropylene than with methanol to form the methoxyl radical required by our hypothetical sequence.

If what we need is a regiospecific C–C bond formation between the benzyl carbon of **I** and the more highly substituted (or more positive) carbon of the olefin, there is another coupling process one can think of. This is the [2+2] photocycloaddition of **I** and **II** as a first step (see Scheme 40.3). This process

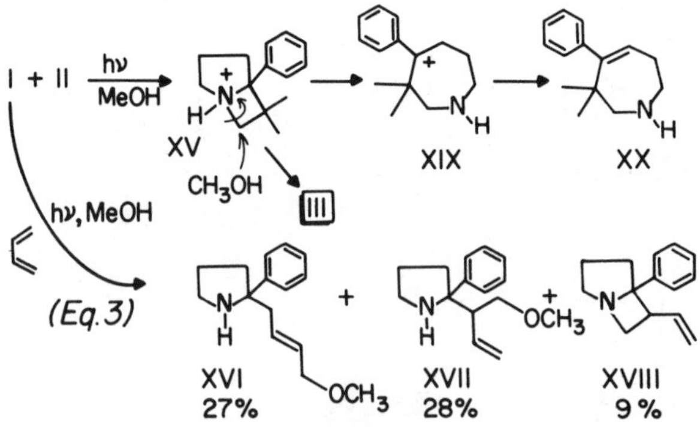

SCHEME 40.3

finds precedent in similar cycloadditions of neutral imines[5] and ketones.[6] Besides, these additions are reportedly regiospecific in the desired direction.[7] Cation **XV** would then be produced.

Cation **XV** opens a number of routes because the quaternary nitrogen atom elicits a high degree of electrophilicity in all its vicinal carbons. Of the three α carbons in **XV,** the methylene carbon of the cyclobutane portion of the molecule should be the more favorable site for bimolecular nucleophilic attack by the methanol solvent, owing to the release of ring strain. This attack would directly yield the observed products **III** and **XII.** Supporting evidence for the existence of **XV** is found in the isolation of **XVIII** [Eq. (3) of Scheme 40.3] during the photolysis of **I** and butadiene in methanol.[1]

The obvious control experiment to test the [2+2] cycloaddition hypothesis is simply to take **XVIII**, protonate it with perchloric acid, and add methanol. This was, in fact, done by the authors of the original investigation[1] with the result that quaternized **XVIII** was stubbornly stable, even in refluxing methanol for an indefinite period of time. This observation is not devoid of precedent.[8]

In addition, polar media such as methanol should favor S_N1 mechanisms where the disconnection of the central or *zero* C–C bond of the bicyclo[3.2.0]heptane structure **XV** should be the more likely process, owing to the stability of the benzylic tertiary carbenium ion **XIX** produced there,[9] and the decrease in ring strain. As a consequence, compounds such as **XX** should have been obtained, something contrary to actual experimental fact. Finally, the lack of reactivity of **XVIII** in the S_N2 sense and its putative equivalent in our system **I**+**II** (intermediate **XV**) may also be due to the neopentylic character of the involved cyclobutyl methylene.

We seem to have explored all the alleys of this chemical labyrinth. A charade! This is the point when one believes that there ought to be a different way to look at reaction mechanisms. The usual thing is to think of bonding processes as two-electron transfers in concerted or ionic routes, and one electron exchanges in the bonding of free radicals. Neither of these is satisfactory in the present case. A way out may be provided by electron transfer theory.[10] One of its fundamental postulates is that, in chemical reactions, electrons can only be transferred one at a time, and not in pairs as is widely assumed for polar reactions. In fact, if reacting molecules are put together at some minimum distance and there is a favorable potential created by this approach, one electron may be transferred from one to the other. For instance, olefins excited by uv radiation may be oxidized to radical cations such as **XXII** by abstraction of one electron, provided there is an efficient electron acceptor. This acceptor may be an electrophilic species that becomes a radical anion if it is originally neutral, or a free

SCHEME 40.4

radical if it is charged positively at the outset. A single electron transfer, therefore, would occur when **I** and **II** approach each other closely as excited species. The chemical complex formed during this encounter is termed an exciplex.[10,11]

There ought to be a finite period of time during which the two active species (depicted by **XXI** and **XXII** of Scheme 40.4) remain in close contact as an exciplex without actually forming C–C bonds. Then radical cation **XXII,** which is certainly an electrophilic species, would be available for nucleophilic attack by methanol at the least substituted carbon, owing to steric interference of substituents. This would give rise to a new complex (**XXIII**) that is now formed by two closely embraced carbon radicals. Collapse of this complex by radical coupling would be expected to be faster than the cyclopropyl radical rearrangement indicated on Scheme 40.2 and would yield product **III** with the observed regiochemistry.

REFERENCES

2. D. R. Adams and S. P. Bhatnagar, *Synthesis*, 661 (1977).
3. This sequence finds a closely related precedent in the work of A. Ledwith, P. J. Russell, and L. H. Sutcliffe, *Proc. R. Soc. London, Ser. A*, **332,** 151 (1973).
4. J. L. Stavinoha, P. S. Mariano, E. Bay, and A. A. Leone, *Tetrahedron Lett.*, 3455 (1980).
5. A. Howard and T. H. Koch, *J. Am. Chem. Soc.*, **97,** 7288 (1975); P. Margaretha, *Helv. Chim. Acta*, **61,** 1025 (1978).
6. C. Rivas, R. A. Bolivar, and M. Cucarella, *J. Heterocycl. Chem.*, **19,** 529 (1982) and references cited therein.

7. J. S. Swenton and J. A. Hyatt, *J. Am. Chem. Soc.*, **96,** 4879 (1974).
8. N. J. Leonard and D. A. Aurand, *J. Org. Chem.*, **33,** 1322 (1968).
9. D. R. Crist and N. J. Leonard, *Agnew. Chem. Int. Ed. Engl.*, **8,** 962 (1969).
10. N. N. Semenov, *Some Problems in Chemical Kinetics and Reactivity*, Princeton University Press, Princeton, 1958; L. Eberson, *Adv. Phys. Org. Chem.*, **18,** 79 (1982). The interested reader is invited to follow a series of papers on organic reactions that proceed via one-electron transfer that were thought to involve two-electron shifts in, E. C. Ashby, J. N. Argyropoulos, G. R. Meyer, and A. B. Goel, *J. Am. Chem. Soc.*, **104,** 6788 (1982) and preceeding papers.
11. R. S. Davidson, *Molecular Association*, Vol. I, R. Foster, Ed., Academic, London, 1975, 215. A. J. Maroulis, Y. Shigemitsu, and D. R. Arnold, *J. Am. Chem. Soc.*, **100,** 535 (1978).

PROBLEM 41

KO–t-Bu

HO–t-Bu

240°, 245 hr

I

II (11%)

+

III

(87%)

1. A. Nickon, J. L. Lambert, J. E. Oliver, D. F. Covey, and J. Morgan, *J. Am. Chem. Soc.*, **98,** 2593 (1976).

PROBLEM 41
A New Reactive Intermediate: The Homoenolate Anion

This reaction includes a highly unusual epimerization of the C-6 carbon of **I** and a skeletal rearrangement in which the quaternary center of **I** ends up as the bridge C-7 carbon of product **III.** This, in turn, entails the formation of a bond between C-6 and C-2, as well as the rupture of the bond linking C-2 and C-1 in compound **I.**

Before going any further it is important to add that the parent hydrocarbon, camphane, does not undergo the slightest change when subjected to the same reaction conditions, so the homolytic abstraction of hydrogen at C-6 that would furnish free radical intermediates is highly unlikely.

Obviously the ketone fragment makes the difference. Most organic chemists think of enolates whenever they see a ketone being exposed to highly alkaline media. An anionic center at C-3 could well be drawn, but its activity might be limited to intramolecular aldol condensations and proton transfer from the solvent or other C-3s of neighboring, neutral molecules of **I.** Nothing of this leads to the observed products. The alternative enolization towards bridgehead carbon C-1 would be precluded by the anti-Bredt character of the resulting double bond.[2] What then?

The only way out is hinted by the epimerization of the C-6 carbon. This

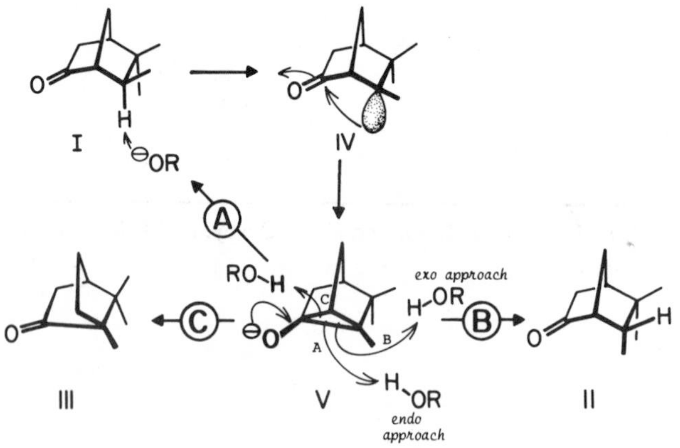

SCHEME 41.1

might entail the tantalizing removal of the proton on this tertiary, seemingly inactive carbon. This being the case, the carbanion thus produced, formally a homoenolate ion,[3] would be highly unstable and would probably act quickly to lower its high energy contents by trapping any electrophilic species of the medium, no matter how feeble. Such an electrophile could be a proton from *tert*-butanol or from another molecule of **I**, or the carbonyl carbon itself. This last alternative, made possible by the rather harsh conditions, is attractive not only because it would yield the desired bond between C-2 and C-6 postulated initially to account for **III**, but the resulting cyclopropane would have three conceivable routes of development: (a) endo protonation to give back the starting ketone (b) exo protonation to afford epimeric compound **II**, and (c) cyclopropane ring disconnection, specifically at the C-2–C-1 bond, to furnish target ketone **III** (see Scheme 41.1).

At this point the reader will probably be haunted by a number of questions: How does it happen that the C-6 hydrogen is selected by the base in favorable competition against hydrogens on C-7 and C-3? What makes it more acidic? How can the distant carbonyl influence the C–H bond elongation? How can a neutral carbonyl survive in this electrophilic form (the enolate is not) under strong basic conditions, to be trapped by the carbanion at C-6? In other words, why is the much more acidic C-3 hydrogen (several orders of magnitude) ignored by the base and the carbonyl not frozen as enolate during the entire 245 h that the reaction lasts?

Some of these questions remain unanswered to this day. Homoenolate anions, however, have been studied since 1962, and presently their structure is understood not as a free β-carbanion such as **IV** but as a much more stable homoanion (**VI**) (see Scheme 41.2).

Careful deuterium exchange and mass spectral studies[1,4] have also shown that proton abstraction in other carbons proximal—but not vicinal—to a ketone is also possible. The intermediacy of rather odd species like **VII**, which feature some of the strain character found in anti-Bredt alkenes and **VIII**, which nec-

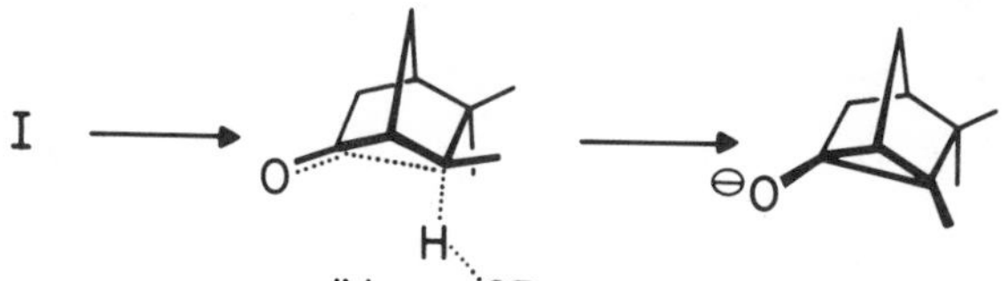

I → IV → V

H, ⊖OR, ⊖O

SCHEME 41.2

III VII VIII

SCHEME 41.3

essarily implies abstraction of a γ proton, have been postulated (see Scheme 41.3).[5,6]

REFERENCES

2. G. L. Buchanan, *Chem. Soc. Rev.*, **3,** 41 (1974); G. Kobrich, *Agnew. Chem. Int. Ed. Engl.*, **12,** 464 (1973).
3. For leading references see A. Nickon and J. Lambert, *J. Am. Chem. Soc.*, **84,** 4604 (1962); A. Nickon, J. Lambert, and J. E. Oliver, *J. Am. Chem. Soc.*, **88,** 2787 (1966); D. Arigoni, *Chem. Commun.*, 597 (1972); J. P. Freeman and J. H. Plonka, *J. Am. Chem. Soc.*, **88,** 3662 (1966); M. Bets and P. Yates, *J. Am. Chem. Soc.*, **92,** 6982 (1970); A. Nickon, D. F. Covey, F. Huang, and Y. Kuo, *J. Am. Chem. Soc.*, **97,** 904 (1975).
4. Homoenolate anions have also been studied by C-13 spectroscopy. See A. L. Johnson, J. B. Stothers, and C. T. Tan, *Can. J. Chem.*, **53,** 212 (1975).
5. The synthetic equivalent of a homoenolate anion has also been developed recently. See D. A. Evans, J. M. Takacs, and K. M. Hurst, *J. Am. Chem. Soc.*, **101,** 371 (1979); O. W. Lever, *Tetrahedron*, **32,** 1943 (1976).
6. Homoenolate anions continue to be of current interest. For recent articles see A. Nickon, H. R. Kwasnik, C. T. Mathew, T. D. Swartz, R. O. Williams, and J. B. DiGiorgio, *J. Org. Chem.*, **43,** 3904 (1978); R. Goswami and D. E. Corcoran, *J. Am. Chem. Soc.*, **105,** 7182 (1983); for a review of homoenolates generated from base treatment of cyclopropanols see N. H. Werstiuk, *Tetrahedron*, **39,** 206 (1983).

PROBLEM 42

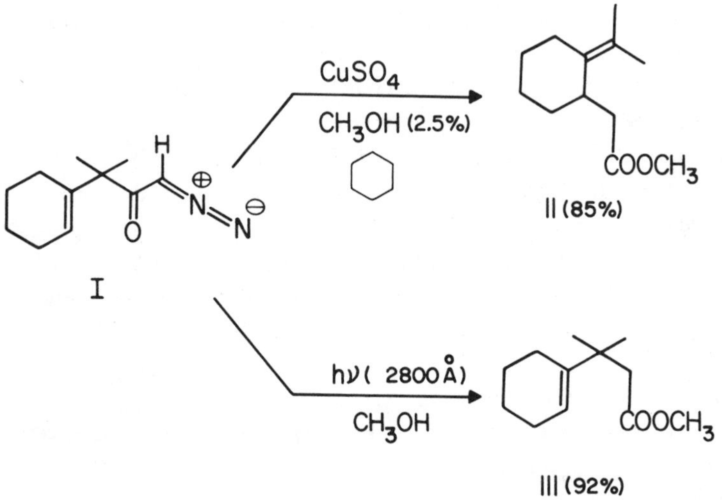

CuSO4
CH3OH (2.5%)
COOCH3
II (85%)
H
O
N
N
I
hν (2800 Å)
CH3OH
COOCH3
III (92%)

1. A. B. Smith, III, J. Chem. Soc. Chem. Commun., 695 (1974); A. B. Smith, III, S. J. Branca, and B. H. Toder, *Tetrahedron Lett.*, 4225 (1975).

PROBLEM 42
First Encounter with Versatile α-Diazoketones: Photolysis versus Copper Catalysis

This first encounter with α-diazoketones in this collection of problems is particularly interesting not only in the sense that it illustrates an unprecedented skeletal rearrangement, but in addition, its discussion leads us to deal with some important aspects of the intricate and versatile chemistry of the diazocarbonyl function.

Let us discuss first the copper sulfate catalyzed reaction. Although transition metal mediated reactions are becoming increasingly popular for organic synthesis,[2] they often involve obscure mechanistic pathways. Among these, few have been subjects of more imaginative speculation than the catalysis by copper salts and copper chelates in the decomposition of the diazo group. This is due in part to the exceedingly fast process involved that precludes precise kinetic studies, and the inability of existing methods to trap the relevant organometallic intermediates.[3] All that is known for sure is that a free carbene—generated by thermolysis or photolysis of a parent diazo compound—and the rather mysterious copper–diazo compound complex, frequently—but not always—feature similar reactivity patterns. This fact and the successful isolation and characterization of some true transition metal carbenes (tungsten, manganese, chromium, iron, and rhenium, but not copper)[4] have provided organic chemists with a relatively safe, but at times questionable solution to this problem.[5] Customarily a copper carbene complex named *copper carbenoid* is attributed to the fuzzily defined and elusive metal organic intermediate.

The reaction plan of this problem includes a reductive positioning of an acetic ester residue on the monosubstituted olefinic carbon of **I** to account for **II.** Conversely, this double bond remains apparently untouched in **III** and the acetic ester shows up now in a neopentyl situation. These overall transformations, therefore, must involve drastic rearrangements of the diazocarbonyl function probably by means of a series of C–C bond forming and breaking steps. Among the C–C bond forming steps there ought to be one whereby the diazo-bearing carbon and the less substituted olefinic carbon become bonded, in order to explain product **II,** as opposed to the process leading to **III,** where this step does not occur.

Now, the starting diazoketone **I** presumably would become the copper complex **IV** after exclusion of molecular nitrogen by the works of the metal salt—

as yet not formally defined, (see Scheme 42.1). If cyclopropanation is the most characteristic process experienced by α-keto carbenoids, there is no particular reason to propose that anything different occurs with **IV,** except perhaps for the considerable ring strain involved.[6] Thus carbocycle **V** would result. By this means, a C–C bond is established between the monosubstituted trigonal endocyclic carbon and the diazo-bearing carbon, the requisite indicated previously. In species **V** a fragmentation process in which this C–C bond is maintained is conceivable. In fact, a retro [2+2] cycloaddition in the way indicated on Scheme 42.1 would yield ketene derivative **VI,** which is a direct precursor of the final methyl ester **II** by intervention of methanol solvent on its electrophilic central carbon atom.

Alternatively, a structure much less demanding than **V** such as **VII** could also result by intramolecular monoalkylation of the copper carbenoid **IV.** This intermediate contains an electron deficient center two atoms removed from a carbonyl function. This is the combination required for acidic-type cleavages such as in β-dicarbonyls and β-keto esters. One would expect, therefore, the occurrence of the nucleophilic attack–bond fragmentation sequence represented in structure **VIII** to yield **II.**

Both **V** and **VII** are highly unstable species, as is copper carbenoid **IV.** It is conceivable that both evolutionary alternatives follow downhill energy profiles.[7] Whatever the particular mechanism, the transformation of **I** into **II** has been termed the vinylogous Wolff rearrangement,[8] since it was taken as a homolog of the classical Wolff transposition.

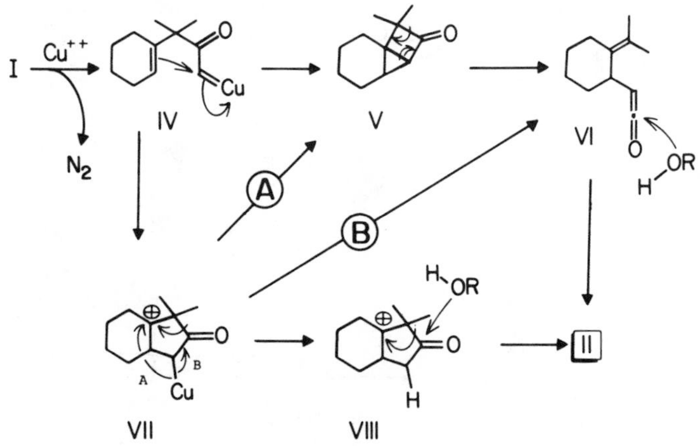

SCHEME 42.1

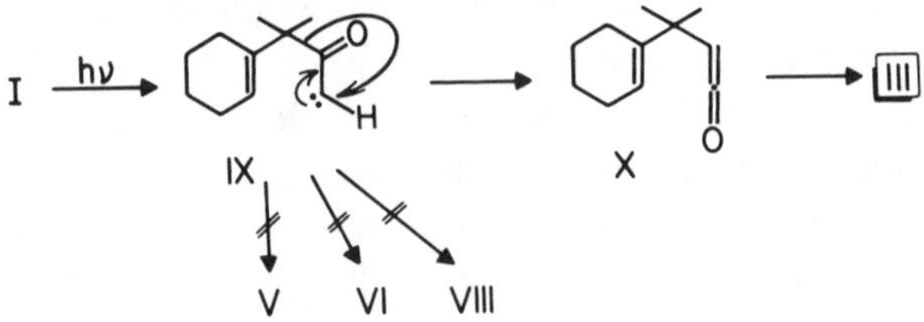

SCHEME 42.2

An attempt to differentiate between these two alternatives has been made by labeling experiments with deuterium at those positions occupied by the *gem*-dimethyls in **I.** Despite the author's claims in favor of bicyclobutanone **V,** these experiments, in our modest opinion, fail to give an ambiguous answer.[9]

The divergent behavior of this photochemical reaction is an illustration of those instances where a *free carbene* and a copper carbenoid show different

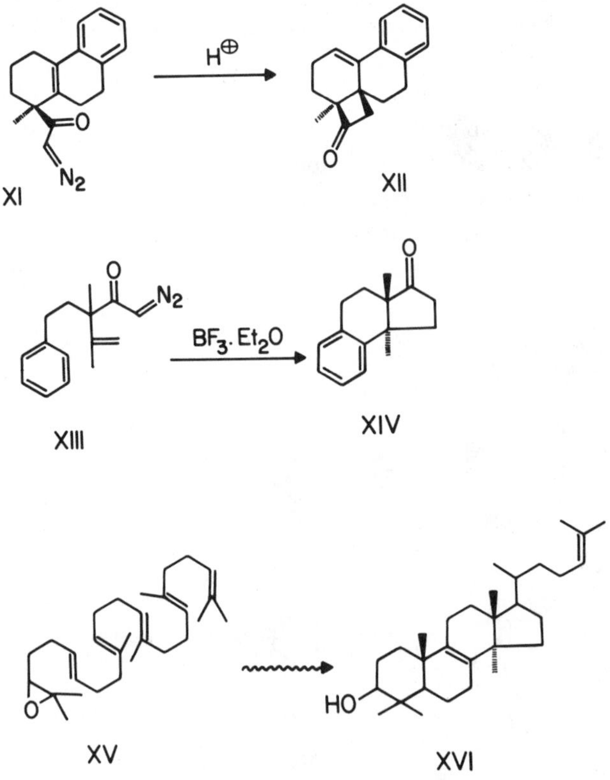

SCHEME 42.3

chemical behavior.[10] The photolysis of **I** also involves skeletal reorganization—the real Wolff rearrangement—where the ketene **X** is formed without intervention of the C=C bond. Structure **X** is then transformed into the ester as previously mentioned (see Scheme 42.2).

Interestingly, β,γ-unsaturated α-diazoketones are also sources of cyclobutanones when they are exposed to protic acid.[1] For example, compound **XI** furnished **XII** in high yield upon contact with concentrated sulfuric acid (see Scheme 42.3). In a conceptually analogous reaction, β,γ-unsaturated α-diazoketones proved to be useful in the construction of cyclopentanones **XIV**[11] in a polyolefinic cationic cyclization process reminiscent of the mechanism by which plants in nature build their polycyclic triterpenoid metabolites from squalene, that is, **XV** → **XVI**.[12]

In Scheme 42.4 the diazoketone operates in a third mode. Here the diazo compound acts as an electron donor, the dominant process when it is exposed to protic or Lewis acids. The resulting intermediate **XVIII** then becomes

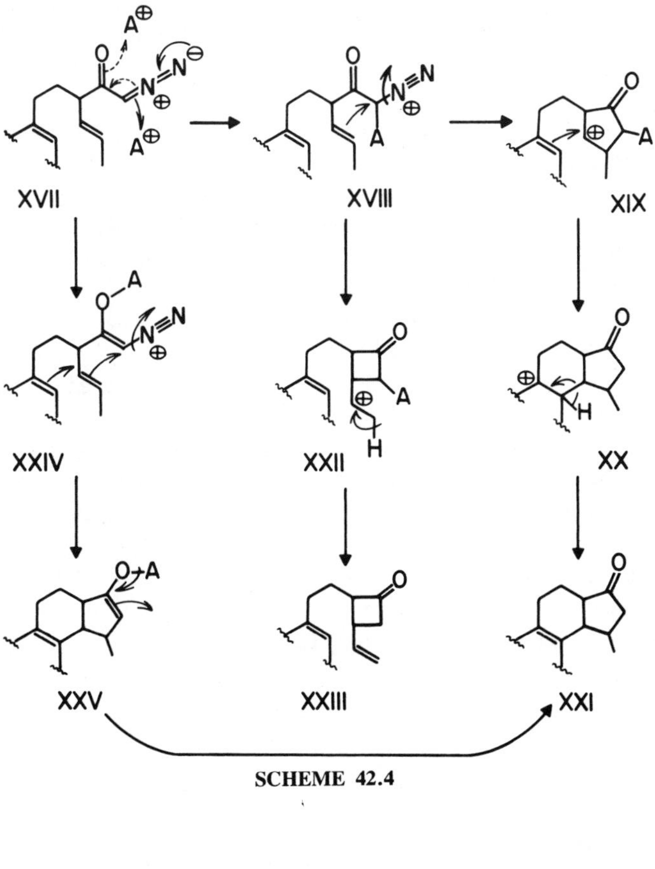

XVII XVIII XIX XXIV XXII XX XXV XXIII XXI

SCHEME 42.4

strongly electrophilic at the diazo carbon. The proximal alkene moiety then may interact with this center in two directions giving rise to cyclopentanones (**XXI**) or cyclobutanones (**XXIII**) from the same common intermediate.

In short, β,γ-unsaturated diazoketones can lead to at least four divergent processes on which almost complete chemical control may be exerted by judicious manipulation of reaction conditions and the structure of substrate compounds. It is advisable that the reader bear these modes in mind for future problems that include this tremendously versatile functionality.

REFERENCES

2. See, for example, *Transition Metal Organometallics in Organic Synthesis*, Vols. I and II, H. Alper, Ed., Academic, New York, 1976; G. H. Posner, *An Introduction to Synthesis using Organocopper Reagents*, Wiley-Interscience, New York, 1980.
3. This forces the researcher to study the mechanism only by reaction product analysis under controlled conditions. For a specific example on diazocarbonyl chemistry see M. E. Alonso, A. Morales, and A. W. Chitty, *J. Org. Chem.*, **47,** 3747 (1982); M. E. Alonso and M.-C. Garcia, *J. Org. Chem.*, **50,** 988 (1985).
4. See, for example, E. O. Fischer, U. Schubert, and H. Fischer *Pure Appl. Chem.*, **50,** 857 (1978); M. Brookhart and J. R. Tucker, *J. Am. Chem. Soc.*, **103,** 979 (1981) and references cited therein.
5. See, among many other hypotheses the one presented by M. P. Doyle, J. H. Griffin, V. Bagheri, and R. L. Dorow, *Organometallics*, **3,** 53 (1984).
6. Highly strained cyclobutanones have been known for years. See W. v. E. Doering and M. Pomerantz, *Tetrahedron Lett.*, 961 (1964).
7. For a similar rearrangement see A. L. Wilds, R. L. von Tebra, and N. F. Woolsey, *J. Org. Chem.*, **34,** 2404 (1969).
8. A. B. Smith, III, B. H. Toder, and S. J. Branca, *J. Am. Chem. Soc.*, **98,** 7456 (1976).
9. The interested readers are invited to inquire into this incomplete problem and decide themselves if the evidence presented is sufficiently unambiguous. See J. P. Lokensgards, J. O'Dea, and E. A. Hill, *J. Org. Chem.*, **39,** 3355 (1974). See also reference (8).
10. U. R. Ghatak and B. Sanyal, *J. Chem. Soc. Chem. Commun.*, 876 (1974).
11. A. B. Smith, III, and R. K. Dieter, *J. Org. Chem.*, **42,** 396 (1977).
12. E. E. van Tamelen, *Acc. Chem. Res.*, **1,** 111 (1968).

PROBLEM 43

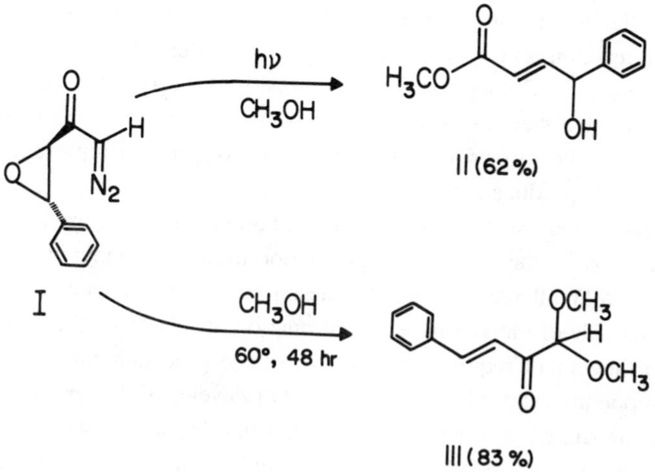

I

$h\nu$

CH_3OH

II (62%)

CH_3OH

60°, 48 hr

III (83%)

1. N. F. Woolsey and M. H. Khalil, *J. Org. Chem.*, **40,** 3521 (1975).

PROBLEM 43
Second Encounter with Versatile α-Diazoketones: Photolysis versus Reaction with Methanol

The convoluted chemistry of diazoketones has already become apparent in a previous problem. That, however, was just a small sample of what these synthons can do. This example embodies some further aspects of importance to their challenging chemistry.

With diazoketones that include additional functionalization on the side of the carbonyl carbon opposite the diazo function, complications arise in part because of the interaction of these functions with the diazo-bearing carbon. That the present reaction is one such case is evidenced by the reductive disappearance of the epoxide unit. Compound **I,** in fact, undergoes an overall reduction during its conversion to either **II** or **III,** although no reducing agents are included here. This should not be surprising since molecular nitrogen that represents the oxidized species is produced in the reaction.

The ester group in **II** is suggestive—although it is not a proof—of the intermediacy of a ketene, and ketene production in diazocarbonyl chemistry usually implies a Wolff rearrangement. The construction of a three-carbon chain on the other side of the ketone is a confirmation of this prediction. In turn, the Wolff rearrangement requires an α-keto carbene precursor that is the fate of diazo compounds exposed to ultraviolet light (wavelength lower than 3200 Å). All this is translated into the mechanism depicted in Scheme 43.1.

There are those who feel disenchanted by simple answers, who enjoy the

hν
(A)
R =
I
IV
V
II
CH_3OH
H_3CO
OH
$HOCH_3$

SCHEME 43.1

challenge of the abstruse. They should now rejoice with the complications included in the alternative pathway to compound **II** now described. It is essentially the transformation of a ketone into an ester while all the substituents are placed in one corner of the molecule. This operation makes use of a Favorskii-type rearrangement.

The key to Favorskii transpositions is a transient cyclopropanone. In this case, the cyclopropanone would be obtained by means of an odd, although not unprecedented, route based on the interaction of carbene and epoxide as indicated in Scheme 43.2. In fact, we observed for the first time some years ago that diazoketone **VI** was converted to enone **IX** upon exposure to copper bronze in refluxing benzene, presumably via oxetonium ion **VIII.**[2]

Translation of these results into compound **I** leads to structure **X.** Unraveling of the strained zwitterion **XI** derived from this would yield keto aldehyde **XII,** a structure that plays a central role in the various possible reaction mechanisms that branch off from the starting material **I.** Furthermore, under photolytic conditions, some alkenes react with carbonyl compounds to form four-membered cyclic ethers, namely, oxetanes, by way of a [2+2] cycloaddition reaction[3] known as the Paterno–Buchi process.[4] Such a reaction would be all that is necessary to convert **XII** into the bicyclic cyclopropanone **XIII** required for the Favorskii-type rearrangement (see Scheme 42.3). Splitting by methanol attack would directly yield compound **II.**

Although the beauty of this approach is undeniable it is incorrect, since the irradiation of pure **XII** synthesized independently leads only to a tar.[1]

Let us now put into practice other patterns of chemical aesthetics in trying to account for **III.** It should be clear by now that **III** is just a dimethyl acetal derivative of **XII.** Then the problem would be solved if we manage to convert

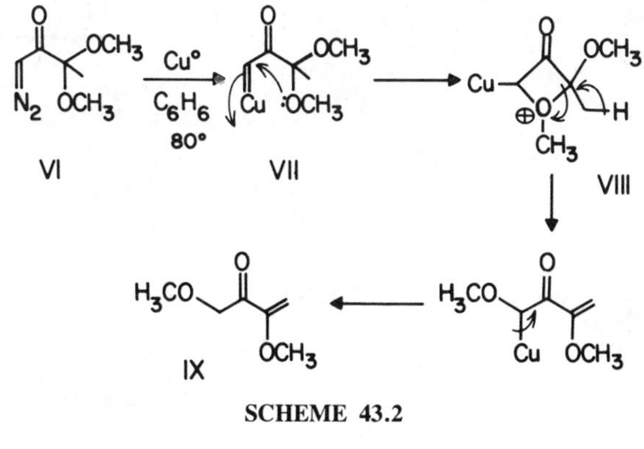

SCHEME 43.2

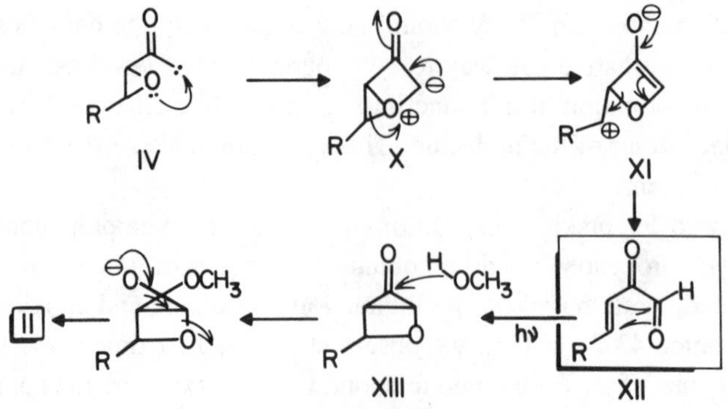

SCHEME 43.3

I into **XII,** necessarily by routes different from those portrayed in Schemes 43.1 and 43.3 since the formation of a carbene in boiling methanol would be superseded by other more favorable processes.

The problem may be restated by saying that the oxygen atom in the oxirane ring must become bonded somehow to the carbon α to the carbonyl in order to

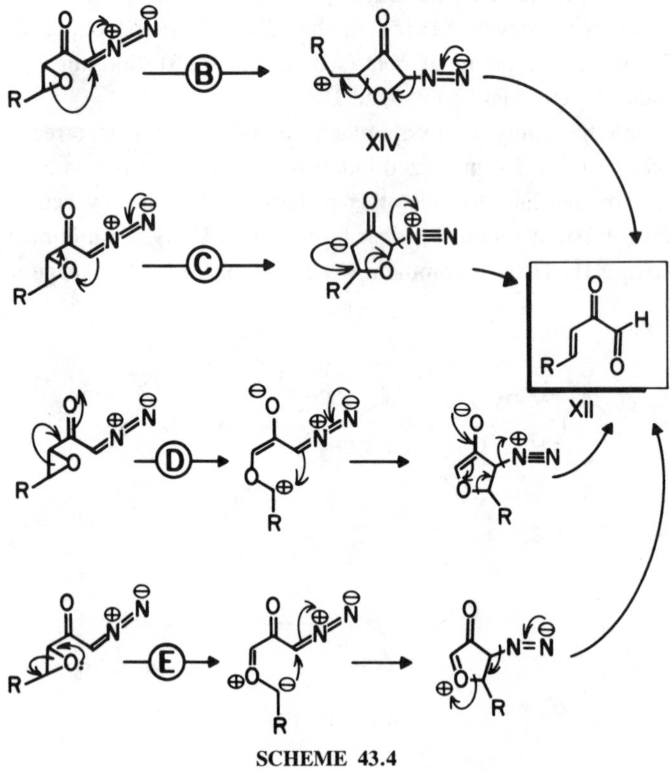

SCHEME 43.4

provide for the aldehydo function in **XII.** This task may be achieved in principle by the intermediacy of cation **XIV** (route **B** of Scheme 43.4) that is formally equivalent to **XI.** Its construction might find justification in the electrophilic character of the diazo carbon and the stability of the benzylic cation. Conversely, the breaking of the other C–O bond of the epoxide (route **C**) in the opposite sense (oxygen is the receptor of bonding electrons from the diazo function) is also conceivable. It would be the consequence of the nucleophilic character of the diazo carbon and the stability of the α-keto anion. As we see, the diazo carbon may behave electro- or nucleophilically, that is, in an ambiphilic fashion depending on the workers' convenience.

Still two other possibilities (routes **D** and **E**) are open to compound **I.** These make use of the rupture of the third bond of the oxirane function (the C–C bond) and the ambiphilic nature of the versatile diazo carbon.

Although routes **D** and **E** can be regarded as unlikely on the grounds that placing a deuterium atom on the diazo carbon in **I** yields **III** with deuterium at the acetal carbon, a final choice between routes **B** and **C** cannot be made with presently available data.

REFERENCES

2. M. E. Alonso, Ph.D. dissertation, Indiana University, 1974. See also E. Wenkert, *Heterocycles*, **14,** 1703 (1980).
3. See, for example, C. Rivas, R. A. Bolivar, and M. Cucarella, *J. Heterocycl. Chem.*, **19,** 529 (1982).
4. A. Padwa, *Organic Photochemistry*, Vol. II, Dekker, New York, 1969.

PROBLEM 44

F_3CCOOH

Et_2O

0°, 3 MIN

I

II

1. R. A. Volkmann, G. C. Andrews, and W. S. Johnson, *J. Am. Chem. Soc.*, **97**, 4777 (1975).

PROBLEM 44
[3+2] Cycloaddition of an Alkyne with Bridgehead Nonclassical Carbocation Mediator

Nothing really arcane prevents the rationalization of this isomerization, as long as a reaction mechanism is deemed simply a reorganization of little sticks. Some unusual features will nevertheless become apparent in working the possible mechanistic paths in some detail.

Bond counting indicates that two C–C bonds are being formed in the process. Following the fate of the six-carbon alkynyl chain easily reveals that these two C–C bonds are those between the *sp* carbons and the two nonadjacent carbons of the cyclopentenol ring. These two nonvicinal carbons appear conveniently activated (carbinol and alkenyl) for this purpose (see structure **III**).

If one of the chaining points is a carbinol, it is logical to expect that C–C bond formation at this point might occur by attack of the alkyne function acting as a donor of electron density. Consequently, the second C–C bond must result from the nucleophilic attack of the cyclic olefin onto the now positively charged ex-alkyne. The two processes require rather high energies of activation. In order for the reaction to go, a strong driving force such as the formation of a very low electron density center must be present.

This preliminary rationale is accommodated in a mechanistic model that begins with the departure of alcohol as water prompted by the strongly acidic medium, and the trapping of the resulting allyl cation **IV** with the acetylene function. Although vinyl cation **V** would have been a proposition open to public ridicule years ago, today it is a well established reactive intermediate.[2] Its actual participation in this reaction was seen by the isolation of ketone **VIII** from the reaction, which may have been derived from the attack of water on this cation (**Va**) (see Scheme 44.1). Also, ample precedent exists for the attack of alkynes on cationic centers.[3]

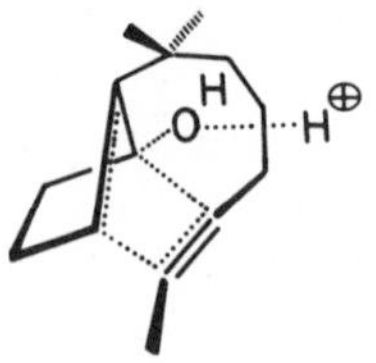

III

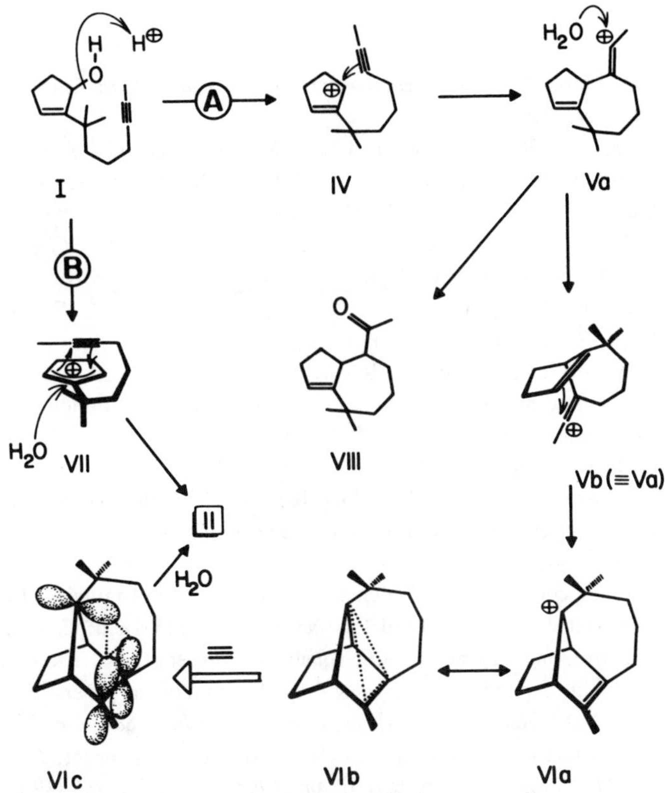

SCHEME 44.1

Structure **V** would be ideally constituted (see **Vb**) for the construction of the second C–C bond required by structure **II.** One major roadblock, however, is found at this point. It is the formation of a very unusual bridgehead carbocation (**VIa**) that violates Bredt's rule. Nevertheless, carbenium ions at bridgeheads of polycyclic structures are well accepted reaction intermediates.[4] Furthermore, structure **VI,** being an *anti* norbornenyl cation, benefits from the additional potential stabilization of its nonclassical carbenium ion character.[5] It would not sit around long before water converted it to the longifolene precursor (**II**).

The classical concerted cycloaddition of the acetylene group on the allyl cation **VII** (route **B**) would account for product **II,** were it not for the forbidden nature of this transition according to Woodward–Hoffman symmetry rules.[6]

REFERENCES

2. See, for example, N. Issacs, *Reactive Intermediates in Organic Chemistry*, Wiley, New York, 1979, p. 120.
3. See, for instance, P. E. Peterson and R. J. Kamat, *J. Am. Chem. Soc.*, **91,** 4521 (1969) and references cited therein. W. S. Johnson, M. B. Gravestock, R. J. Parry, R. F. Meyers, T. A. Bryson, and D. H. Miles, *J. Am. Chem. Soc.*, **93,** 4330 (1971); W. S. Johnson, M. B. Gravestock, and B. E. McCarry, *J. Am. Chem. Soc.*, **93,** 4332 (1971). Cationic olefin cyclizations have been studied extensively. For reviews see W. S. Johnson, *Angew. Chem. Int. Ed. Engl.*, **15,** 9 (1976); E. E. VanTamelen, *Acc. Chem. Res.*, **8,** 152 (1975). For an interesting recent olefin cyclization rearrangement see W. W. Epstein, J. R. Grua, and D. Gregonis, *J. Org. Chem.*, **47,** 1128 (1982). Alkyne alkylations by bimolecular displacement on low electron density neutral compounds also occurs. See F. Marcuzzi, G. Modena, and G. Melloni, *J. Org. Chem.*, **47,** 4577 (1982).
4. The natural product longifolene to which **II** is closely related, is specially prone to yielding the unorthodox bridgehead cation. See among others, S. Dev, *Acc. Chem. Res.*, **14,** 82 (1981).
5. This important concept of modern chemistry had a difficult beginning, for its introduction by Saul Winstein and Daniel Trifan in 1949 led to one of the most fascinating, instructive, and lengthy scientific squabbles of recent times. The reader should have a gratifying experience in browsing through the following papers: H. C. Brown, *Chem. Br.*, 199 (1966); G. D. Sargent, *Quart. Rev. Chem. Soc. (London)*, **20,** 301 (1966); S. Winstein, F. Clippinger, R. Howe, and E. Vogelfanger, *J. Am. Chem. Soc.*, **87,** 376 (1965); S. Winstein, *J. Am. Chem. Soc.*, **87,** 381 (1965), and the leading references cited in these articles. The controversy continues to this date, but now in more sophisticated terms. See H. C. Brown, M. Periasami, D. P. Kelly, and J. J. Giansiracusa, *J. Org. Chem.*, **47,** 2089 (1982); G. A. Olah, G. K. S. Prakash, D. Farnum, and T. Clausen, *J. Org. Chem.*, **48,** 2146 (1983).
6. R. B. Woodward and R. Hoffmann, *The Conservation of Orbital Symmetry*, Verlag Chemie/Academic, 1970.

PROBLEM 45

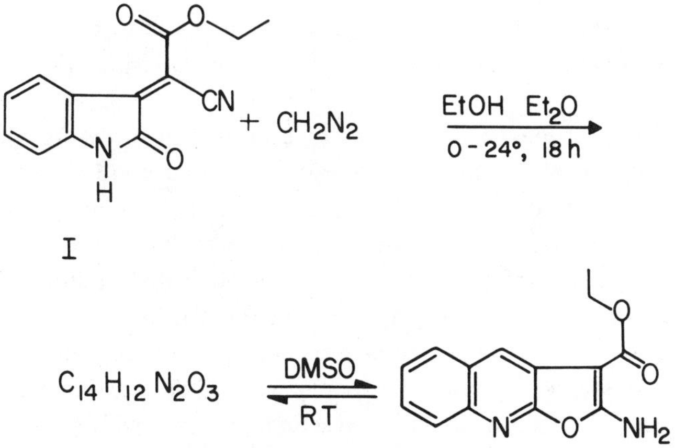

I + CH_2N_2 $\xrightarrow[0-24°,\ 18\ h]{EtOH\ \ Et_2O}$

$C_{14}H_{12}N_2O_3$ (II) $\underset{RT}{\overset{DMSO}{\rightleftharpoons}}$ III

1. G. R. Bennett, R. B. Mason, and M. J. Shapiro, *J. Org. Chem.*, **43,** 4383 (1978).

PROBLEM 45
An Unusual Fate for a Pyrazoline Derived from Diazomethane

In the struggle of solving any problem, things may appear agonizingly obscure, particularly if the task of retrieving the pertinent experimental information is left to one's judgment alone. This problem is presented here to better illustrate the everyday situation of the practicing researcher. Thus, only part of the evidence has been presented by concealing the explicit structure of compound **II.** This would convey the real situation if, say, the structure of **III** became simpler to elucidate than that of **II** with the spectrometers at our disposal. This initial handicap, however, will be compensated for by the description of some additional results.

Comparison of empirical formulas reveals two important features. First, only the methylene of diazomethane is being incorporated in **I** to yield **II,** since there is no other source of methylenes present and the two nitrogen atoms belong to the indolinine ring and the nitrile substituent. This fact alone allows one to rule out the possible contribution of a pyrazoline ring in structure **II.** Although the participation of this ring in an intermediary stage cannot be disregarded, in view of previous experience pertaining to the construction of compound **V** from the closely related oxindole acrylate **IV** (see Scheme 45.1). In addition, one of the most important reactions of diazomethane with olefinic substrates in the absence of light or catalysis by metals or Lewis acids is the 1,3 dipolar cycloaddition that affords pyrazolines.[2]

IV + CH_2N_2 —I→ V —Δ, $C_6H_4(CH_3)_2$, $-N_2$→ VI ($C_{13}H_{13}NO_3$)

I) = Et_2O, 0–20°, 18 h

SCHEME 45.1

It is noteworthy that in the case of **IV** there is a unidirectional, regioselective formation of the pyrazoline ring due to the more electrophilic character of the α carbon of the acrylate fragment than that of the β-methyne. The regioselectivity of this cycloaddition notwithstanding, pyrazolines usually collapse to cyclopropanes by exclusion of molecular nitrogen under the influence of moderate heat, such as in **V** leading to **VI.**

The keen observer may have noticed already that the empirical formula of **VI** is that of **II** minus an HCN fragment. It is also of importance to recognize that cyclopropanation is indeed an excellent method to incorporate a methylene unit into **I.** Then, one is led to believe that **II** ought to be the cyclopropane derivative **VIII,** produced in all probability from pyrazoline intermediate **VII** (see Scheme 45.2).

A nineteenth century chemist might have left things like this. Nowadays, however, spectroscopic methods provide us with such a wealth of data that they prove our mental contrivances wrong once and again. This happens to be the present case, for as we shall see, not only is **VIII** not the correct structure, but **VII** is questionable as well.

To this end we draw the second pending conclusion from the analysis of empirical formulas. Structures **II** and **III** are isomeric, and the equilibrium between these two systems in a highly polar solvent and at a moderate temperature suggests a very close structural relationship and resemblance. Even though the cyclopropane fragment in **VIII** would be highly active in terms of rearrangement as a consequence of having two geminal electron withdrawing substituents,[3] it would also be prone to isomerize by way of a 1,3-sigmatropic rearrangement to the entirely different compounds **IX** and **X** (see Scheme 45.3).

Fragmentation analysis of compound **III,** as shown in Scheme 45.4, allows one to discriminate its composing parts in terms of the starting materials. Thus, the amino furan moiety finds a cursory explanation, and it places the amide and cyanoacetic ester functions as parts of compound **II.**

More importantly, it also becomes apparent that the methylene from di-

I + CH_2N_2 → VII → ($-N_2$) VIII

SCHEME 45.2

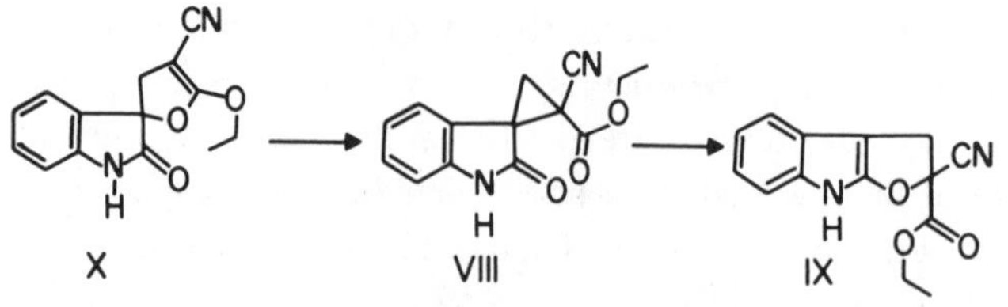

SCHEME 45.3

azomethane has found its way into the pyridine ring of the quinoline structure. The obvious ring expansion that this involves ought to have its precursor in some sort of addition product of diazomethane and **I.** Neither cyclopropane **VIII** nor pyrazoline **VII** seem to fulfill this purpose since their expected bond reorganization would yield seven-membered rings (see Scheme 45.5). What do we do next?

A way out may be found if the cycloaddition of diazomethane to **I** occurs in the reverse direction as represented in **XVII.** Now the newly introduced methylene will be in the appropriate position for ring expansion to a six-membered ring. This requires that the first new C–C bond of **I** with diazomethane be formed at the β carbon of the cyano acrylate group as opposed to **V.** This is possible thanks to the inversion of polarity or umpolung of the double bond due to its conjugation with the nitrile. This effect along with the influence of the ester overrules the polarization pull of the lactam ring.

Scheme 45.6 may not be the only possible mechanistic interpretation that accounts for **II** with these restrictions. If one plays a little with "rejected" intermediates **VII** and **VIII,** one might come up with at least one manner of

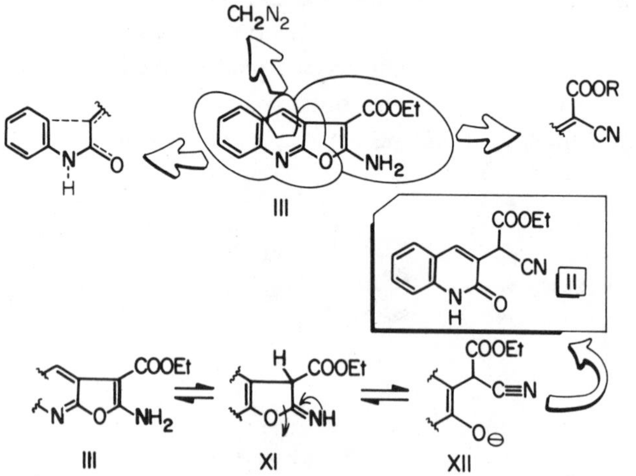

SCHEME 45.4

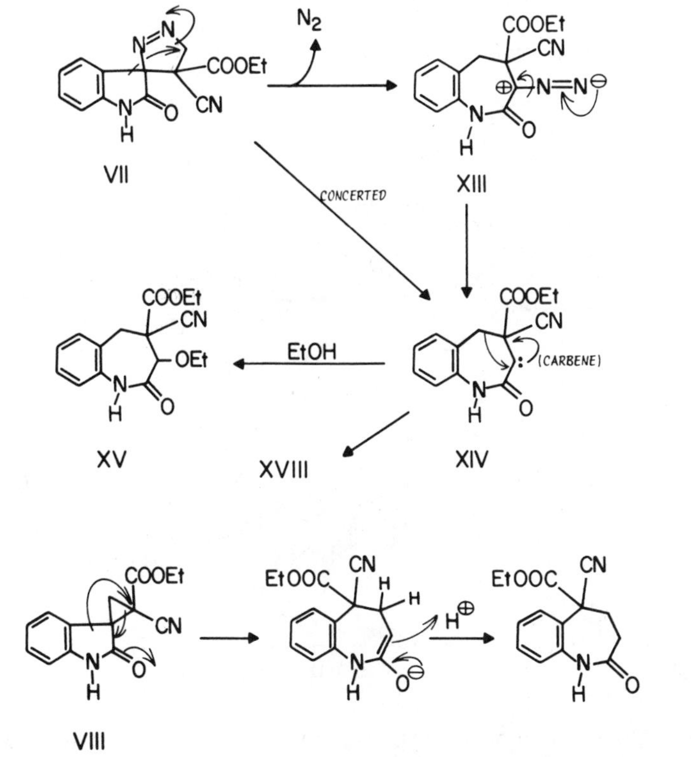

SCHEME 45.5

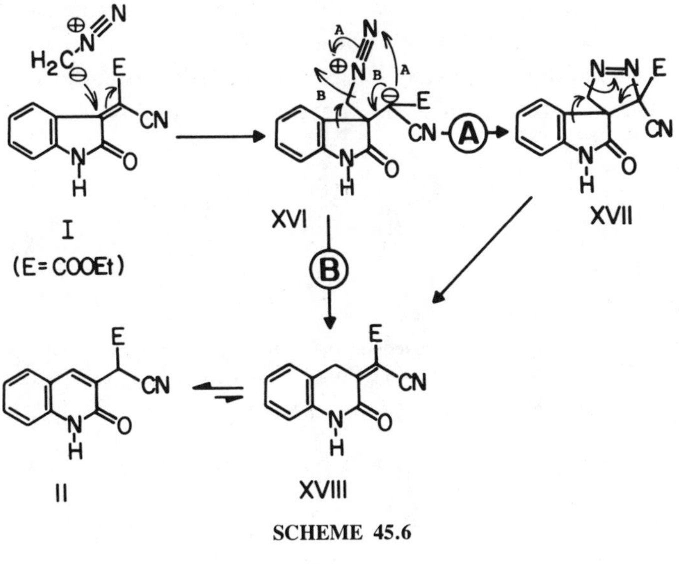

SCHEME 45.6

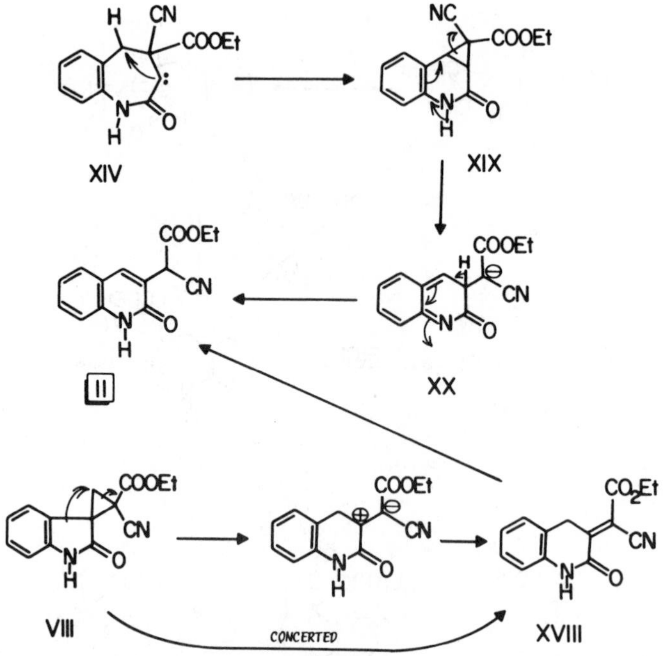

SCHEME 45.7

converting these to compound **II.** The key to these transformations would be carbene **XIV,**[4] which was postulated already in Scheme 45.5, and the strongly electrophilic character of 1-cyano-1-carboethoxycyclopropanes **VII** and **XIX.** Also, the rearrangement of carbene **XIV** to **XVIII,** which is an intermediate precursor of **II,** is conceivable (see Scheme 45.7).

At the present time it cannot be ascertained to what extent this scheme is valid, for there is a paucity of pertinent experimental results.

REFERENCES

2. R. Huisgen, *J. Org. Chem.*, **41,** 403 (1976).
3. See Problems 15, 16, 17, and 24.
4. The ring contractions of benzo fused seven-membered heterocycles have been extensively investigated. See H. C. Van Der Plas, *Ring Transformations of Heterocycles*, Vol. 2, Academic, New York, 1973, pp. 282–321.

PROBLEM 46

HOCl

?

K_2CO_3

CH_3OH-H_2O

KOH

I

II + III + IV

V

1. J. Wolinsky and M. K. Vogel, *J. Org. Chem.*, **42,** 249 (1977).

PROBLEM 46
Vital Importance of Relative Stereochemistry of Substituents for the Reaction Course

Occasionally, even the meticulous analysis of carefully gathered data leads to conclusions that, when confronted with further experimental observations yield a distorted vision of reality. This phenomenon becomes extremely dangerous the moment one ignores it. In this example we ignore it, just to see what happens. Thus the results of the reaction of **I** with hypochlorous acid have been deleted in order for us to figure out by sheer reasoning what might have occurred. We should carefully examine the second step of the sequence. Then we shall amuse ourselves by comparing our theoretical conclusions with the real results of the experiment.

Nothing atypical is customarily found in additions of hypochlorous acid to alkenes, where β-chlorohydrins are the normal products. The subject becomes difficult, however, when substrates prone to labyrinthine chemistry such as β-pinene are used.[2] This subject is not entirely unfamiliar[3] but the original article[1] brings it anew in view of the detailed and instructive study of the stereochemistry.

The first abnormal feature in the initial part of this equation is that four carbons of the molecular backbone of **I** instead of the anticipated two (trigonal carbons) are involved in one way or another as functionalized systems in all products. Thus, some sort of highly active intermediates that deeply affect the carbon skeleton must be formed at some stage. In addition, the presence of oxygen bridges in **II, III,** and **IV** points towards their secondary nature, that is, the primary products of the reaction of **I** with hypochlorous acid are no longer present after treatment of the mixture with base. The stereochemistry of products **II–IV** may reflect, nevertheless, the actual configuration of substituents in the putative precursors as long as the assumption holds that only intramolecular nucleophilic displacements without carbenium ion participation are operative, a perfectly legal assumption in alkaline media. In other words, the detailed study of structures **II, III,** and **IV** may lead not only to the structure of their precursors, but to their absolute stereochemistry as well, hence to the reaction mechanism.

At this point it is important to give consideration to the following features while examining Scheme 46.1:

1. Four active sites:

I ⟹ VI

2. Markovnikov addition of chloride:

Cl–OH → VII → (⊖OH) VIII

II, IV III

3. Common intermediate for **III** and **IV**:

III ⟸ IX ⟹ IV

4. **II** From another precursor:

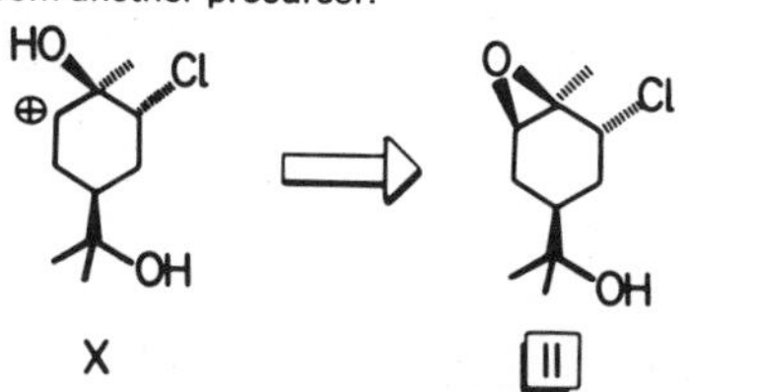

SCHEME 46.1

1. If two of the carbons that underwent change are obviously those pertaining to the double bond of **I**, the two other active sites must be derived from the heterolytic separation of a C–C bond in the strained four-membered ring. The isopropanol chain clearly suggests the position where the positive center to be is located.
2. All three products show signs of a Markovnikov addition of the chlorine atom from the side opposite the *gem*-dimethyl bridge, as is usual in the pinene skeleton. While this is obvious in **II** and **IV**, compound **III** may be conceived as the consequence of the concerted displacement of chloride by the isopropoxy unit approaching from the upper side.
3. Compounds **III** and **IV** are probably related to a common intermediate.

This is inferred from the similar configuration of the epoxide and tertiary alcohol units, and to the configuration of the chlorine bearing carbon. The latter is explicitly displayed in **IV** but is somewhat concealed in **III.** Visibly, the β-oxirane **II** must come from a different precursor, one that features the cyclic carbinol moiety with the opposite configuration.

4. The construction of the oxirane ring in **II** and **III** is more likely the consequence of base treatment of the precursor chlorohydrins rather than ring closure during exposure to hypochlorous acid, owing to the reactivity of epoxides in acidic media.

If, in the presence of hypochlorous acid, α-pinene is to behave as 1-methylcyclohexene does,[4] that is, yielding chiefly Markovnikov addition products

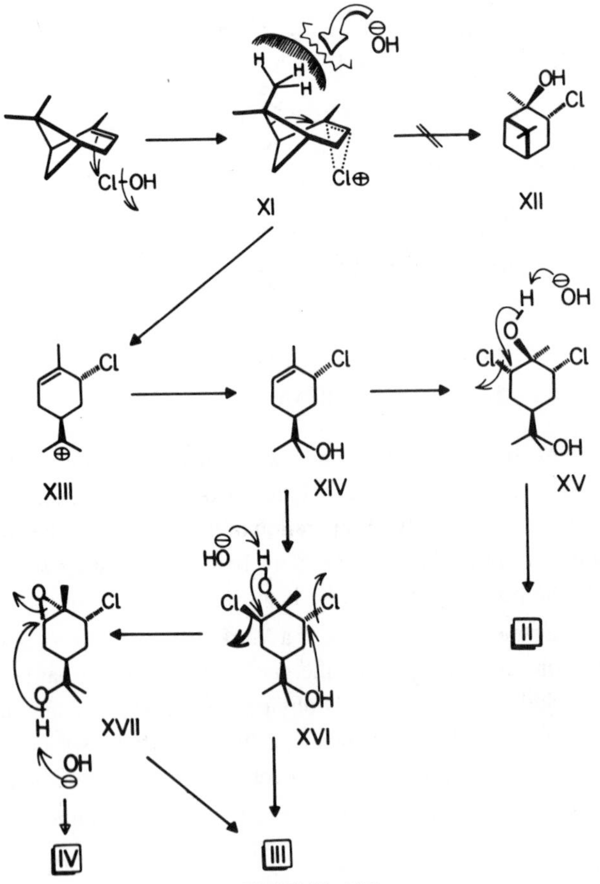

SCHEME 46.2

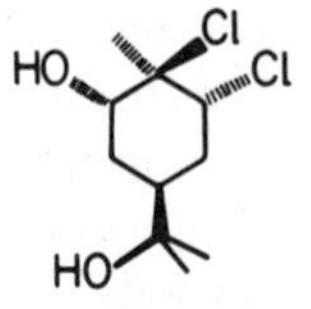

XVIII 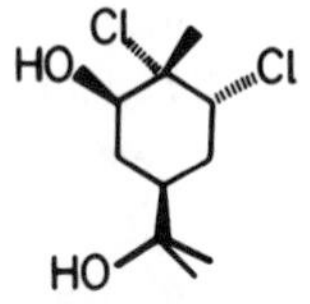 XIX

with the two incoming chloro and hydroxy groups in the trans configuration, but with the proviso of ionic intermediates, then the first stage of the mechanism unfolds clearly. The initially formed chloronium ion either undergoes 1,2 addition to the nonobserved chlorohydrin **XII,** or the four-membered ring unravels with concomitant elimination and hydroxide trapping to **XIV** in what is in essence a 1,6-addition reaction (see Scheme 46.2).

The appearance of a new double bond yields a number of mechanistic possibilities. Most important is the fact that this C=C bond gives access to the introduction of Cl and OH groupings via the chloronium ion pathway with the two different configurations that are needed to account for those precursors of type **IX** and **X,** namely, **XV** and **XVI,** since the stereochemical restrictions imposed by the gem-dimethyl bridge are no longer present.

Conversely, treatment of **XV** with base would trigger the construction of **II** by way of proton abstraction and intramolecular displacement of chloride. At the same time, **XVI** would be the COMMON INTERMEDIATE that was

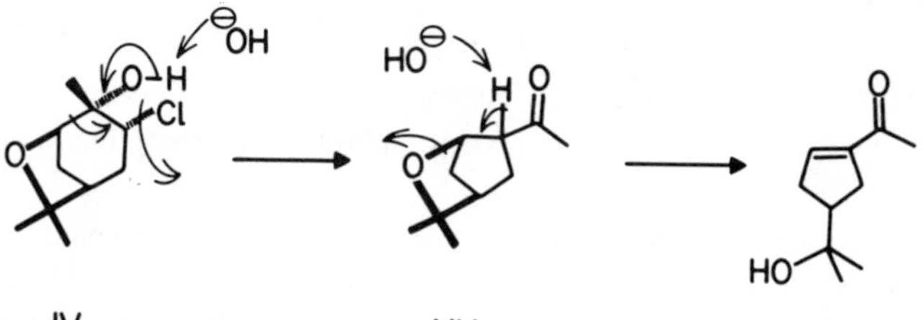

IV XX V

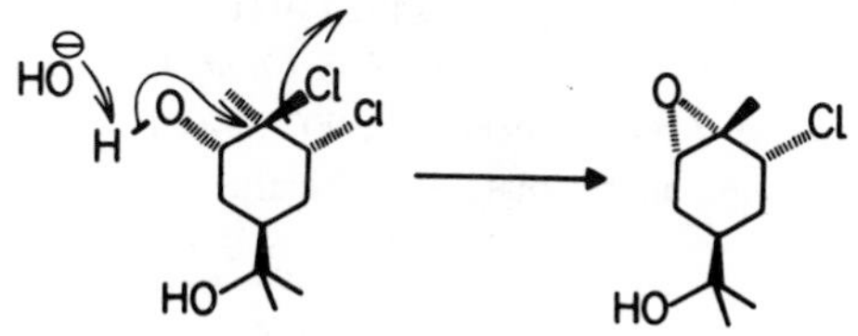

XVIII XVII

SCHEME 46.3

postulated previously, since it contains all the elements required for the production of **III** and **IV.** The latter, though, still needs the intermediacy of epoxide **XVII** in order to facilitate the formation of the five-membered oxygen bridge.

In short, "the production of **II, III,** and **IV** allows one to conclude that only **XV** and **XVI** were formed during the hypochlorous acid step." As it turns out, the scheme described so far is incompatible with actual experimental fact: Our key compound **XVI** was never seen after the first step of the actual experiment. Instead, chlorohydrins **XVIII** and **XIX** (90 and 1%, respectively) were the only compounds characterized, along with 5% of **XV,** our only real contrivance.

The anti-Markovnikov addition that accounts for **XVIII** and **XIX** was attributed to "the inductive effect of the chlorine (in **XIV**) since it is known that addition of hypochlorous acid to allyl chloride gives predominantly 2,3-dichloro-1-propanol."[1,5]

Our previous discussion makes it unnecessary to indicate how compounds **XVIII** and **XIX** yield **XVII,**[6] from which we derive **III** and **IV** as well as **II.** Amusingly enough, when a stronger base (KOH) is used in lieu of potassium carbonate, compound **IV** is converted to cyclopentene **V** in moderate yield. A completely different process reminiscent of a semibenzylic acid rearrangement sets in, in contrast to the epoxidation reaction experienced by chlorohydrin **XVIII** under the same conditions (see Scheme 46.3).

REFERENCES

2. In the mid-1940s there was enough material for a complete review. See J. L. Simonsen, *The Terpenes*, Vol. II, 2nd ed., Cambridge University Press, New York, 1946, pp. 172–176.
3. Professor Wolinsky's work was intended to clean up the stereochemical complications of earlier investigations that use optically impure α-pinene, which, on top of everything else, also contained β-pinene in some cases without the researchers knowing it. See G. Wagner and A. Ginzberg, *Chem. Ber.*, **29,** 886 (1896); G. G. Henderson and J. K Marsh, *J. Chem. Soc.*, **119,** 1492 (1921).
4. H. Bodot, J. Jullien, and M. Mousseron, *Bull. Soc. Chim. Fr.*, 1097 (1958).
5. J. R. Shelton and L. H. Lee, *J. Org. Chem.*, **25,** 907 (1960).
6. Structure **XVII** is the common intermediate we were after!

PROBLEM 47

LiAlH$_4$ (EXCESS), THF, Δ, 56 hr

I

II (5%)

III (11%)

IV (50%)

V (32%)

1. J. Wolinsky, J. H. Thortenson, and M. K. Vogel, *J. Org. Chem.*, **42**, 253 (1977).

PROBLEM 47
Unexpected Confusion during a Lithium Aluminum Hydride Reduction

Since the handling of bicyclic chlorohydrins is still fresh in the reader's mind as a consequence of our lengthy discussion in Problem 46, this is an appropriate occasion for a somewhat more difficult transformation. This example illustrates a common occurrence in chemistry: Even the most trivial reaction may sometimes lead to an unpredictable and surprising result.

Compound **II** was required in moderate quantity and chlorohydrin (**I**) was available in the laboratory from a previous project. Anyone with a standard background of organic synthesis would readily select lithium aluminum hydride (LAH) as the reagent of choice for the apparently simple reduction of the secondary halide. The reaction was in fact run but something funny happened on the way to compound **II.** Three other major products appeared, let alone the fact that **II** was obtained only in 5% yield.

Besides the migration of the oxygen bridge, there are a few additional puzzling facts: Products **III** and **IV** are epimeric, not only at the carbinol carbon, but also at the neighboring methyl-bearing methyne. This might well imply that the carbinol is transformed to a ketone at some stage. This means hydride transfer from the organic skeleton (oxidation) or C–C bond rupture. The second proposition is indeed more likely since it is always difficult to justify an oxidation in the presence of LAH. Whatever the process, this putative ketone would then undergo reduction by way of equatorial attack of LAH to give the axial carbinol **IV.** Also, structure **V** must have been produced from the equivalent of a double oxygen migration.

At the same time, all these processes might be prompted by the shift of the oxygen bridge, whereupon room is left around the HO–C–C(Me) unit for further operations. Such a shift would be driven by a still unclear ejection of chloride anion without the normal hydride transfer to the chlorine-bearing carbon. Curiously, there seems to be a natural propensity of compound **I** towards such a bizarre change since just heating it at 100°C or above causes its equilibration with pinol chlorohydrin **VI.**

Some sort of aluminum-containing species must be involved in this case because the starting material does not undergo any change upon reflux in THF. The assistance by aluminum may be fashioned in several ways. Whatever these may be, they all should entail the separation of a C–Cl bond since the low

I → VI (100°)

SCHEME 47.1

electron density that would presumably develop at the carbon atom would provide a driving pull for the migration of the oxygen bridge. Structure **VIII,** which would be derived from the textbook reaction of an alcohol and LAH, appears to us to be particularly appealing as a starting point, not only because of the comfortable six-membered ring that would be structured, but also because the progressive deformations—geometric and electronic—of this transient ring would lead to the key structure **IX** without having to resort to other sophisticated intermediates (see Scheme 47.2). By this means, the development of a full positive charge that should be unlikely in the presence of excess metal hydride is conveniently avoided.

In addition, intermediate **IX** would be ideally suited to afford all the observed products, by way of four divergent routes:

1. *Route A* of Scheme 47.1: Either **VIII** or **IX** would contemplate the intermolecular transfer of hydride of a second mole of LAH from the exo side, thus furnishing the minor product **II.** Such a bimolecular process, however, would be no competition against the three intramolecular processes that follow.
2. *Route B:* Transfer of hydride from the endo side of the molecule to the tertiary carbon with concurrent displacement of the oxygen bridge. This operation would account for the configuration of compound **III.**
3. *Route C:* This route entails a similar attack by those electrons ''liberated'' during the rupture of the oxygen–aluminum bond. The resulting epoxide **XI** would be further reduced, receiving the hydride at the expected less substituted carbon to give **V** in the right configuration.
4. *Route D:* This route involves the ketone-producing sequence suggested earlier. If a hydride ion is to be shifted from the carbinol carbon in a 1,2 fashion to allow for the consolidation of the ketone,[2] a late transition state, namely, the development of more carbenium ion character of

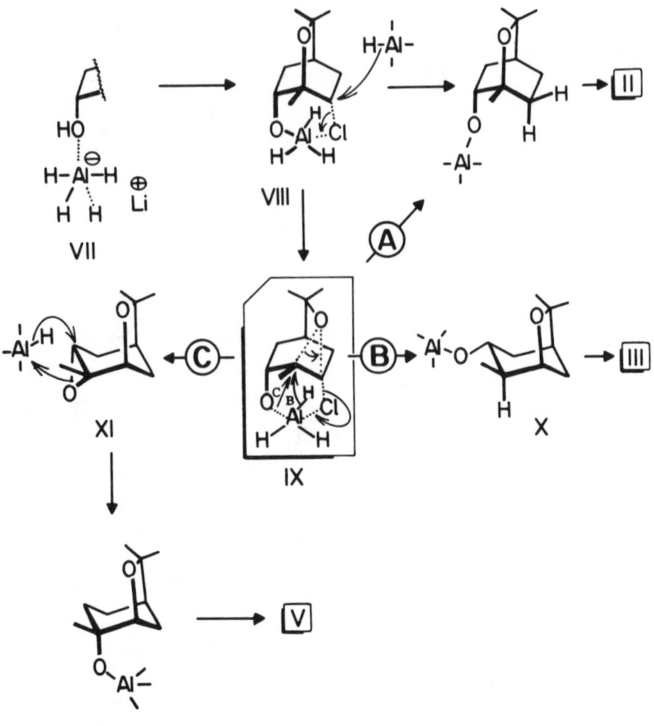

SCHEME 47.2

the bridgehead carbon, should be required for an effective interaction of the orbitals involved.

Although the limited conversion of pinol chlorohydrin **VI** to ketone **XIV** by treatment with only one equivalent of LAH lends support to this idea,[1] **VI** may not be the most appropriate model, because it embodies a more favorable orbital alignment for a concerted 1,2-hydride shift and ejection of the chloride ion. One could argue that compound **I** might isomerize first to **XVI.** However, this proposition would force into existence the rather odd intermediate **XV** (see Scheme 47.3).[3]

Any of these hydride transfer processes yields the observed configuration of the methyl-bearing methyne. Reduction of the ketone through the more favorable equatorial approach finishes the sequence that leads to **IV.**[4]

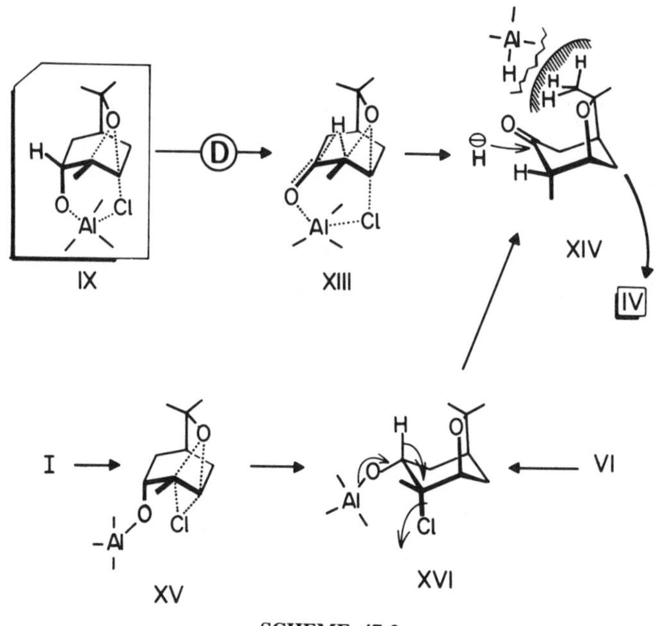

SCHEME 47.3

REFERENCES

2. The reader may try the alternative based on C–C bond breakage. It leads to interesting though unobserved dihydropyranyl structures.
3. The intermediacy of such species would nevertheless be instrumental in accounting for the thermal isomerization of **I** into **IV** mentioned earlier.
4. The occurrence of one-electron transfer in the reduction of organic halides by LAH has been reported recently. Thus a mechanism that includes this option must also be taken into account. See E. C. Ashby, R. N. DePriest, A. B. Goel, B. Wenderoth, and T. N. Pham, *J. Org. Chem.*, **49,** 3545 (1984). See also Problem 40.

PROBLEM 48

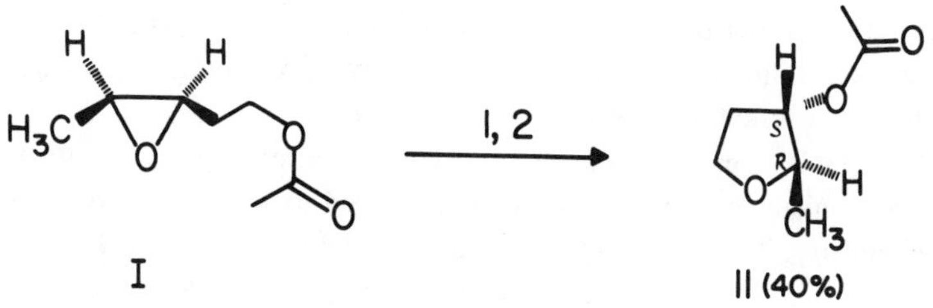

I

II (40%)

$1 = BF_3.Et_2O, Et_2O, 20°, 2.5$ h

$2 = K_2CO_3, H_2O$, 15 min.

1. J. M. Coxon, M. P. Hartshorm, and W. H. Swallow, *J. Org. Chem.*, **39,** 1142 (1974).

PROBLEM 48
Neighboring Group Participation of an Acetoxy Group, the Unexpected Way

Most organic chemists are aware of the notable consequences of subtle changes in the molecular structure of a given compound seen in the mechanism of reactions emanating from it. This is particularly true when these modifications are introduced at atoms near or at polar substituents. An increase in the number of methylenes in a side chain, however, does not fall into this category; it is usually associated with an increase in the size of a ring or other predictable, unexciting phenomena. This problem will remind us that this is not always the case, because it illustrates a most dramatic change of results when just one more methylene is inserted in the acetate-bearing chain.

The notable construction of **II** from **I** may not involve much difficulty as far as mechanisms go, but care must be exercised to account for the observed stereochemistry. To make things more edifying, let us assume that **I** is optically active, with C-3 (*R*) and C-4 (*S*) configurations.

Boron trifluoride is an excellent Lewis acid catalyst to induce the ionic ring opening of oxiranes—among other oxygen containing rings. If a carbocation is generated during this ring rupture in **I,** it should be very short lived in view of the stereospecificity observed. This would also be supported by the fact that the *trans* epimer of **I** yields mostly *cis*-**II.**[1] Indeed, the life span of any of the two secondary carbenium ions that would result from the splitting of the oxirane unit is expected to be considerably affected by the neighboring group participation effect of the acetoxy group. This would also introduce some rigidity to the system, thus favoring stereocontrol. Consequently, stable six- and seven-membered cyclic dioxolenium ions **III** (Scheme 48.1) and **VIII** (Scheme 48.2) would be formed.[2] Both entail inversion of configuration of the involved carbons C-3 and C-4.

Quick-thinking readers may have already grasped the course of action to follow in order to build a THF system with the desired 2,3-substitution from this intermediate. This would be done by drawing structure **III** in its three-dimensional form **IVa**. This would be followed by attack of the oxymethine fragment on C-1, then skipping the rest of this discussion and tackling the next problem. Those who think more slowly—but perhaps more carefully—will soon discover, however, that the product of this process is indeed a THF structure *resembling* **III** but *not identical* to **II,** because the methyl and acetoxy groups

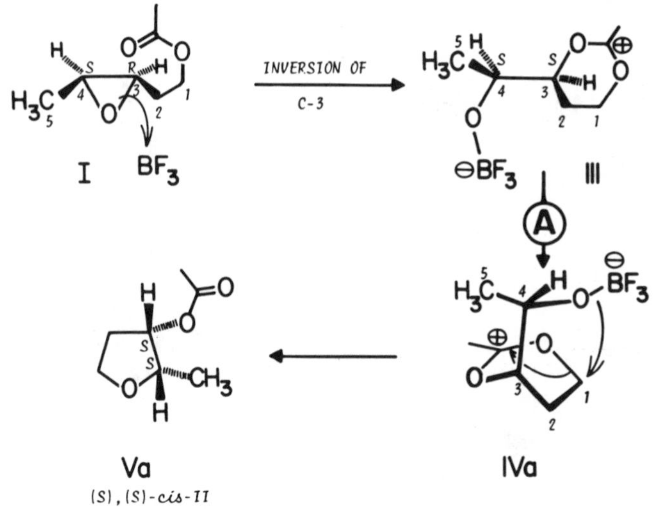

SCHEME 48.1

are *cis* to one another. Therefore, the route that furnishes **II** must be more complicated than what is shown in Scheme 48.1

Structure **IVa** may also be drawn as **IVb** (Scheme 48.2) with the O–B unit sitting just above the empty orbital that constitutes the carbenium ion. The interaction of these two sites would afford the curious ortho ester **VI,** which, having three grossly equivalent C–O bonds present, gives ample room to operate with it. Only two C–O bond ruptures represent forward reactions, however. In turn, while route **C** leads to a structure several parsecs[3] removed from **II,**[4] the bond separation included in route **D** affords a second dioxolenium ion **VII** where *trans*-**II,** namely, **VIIIa**, results after inversion of configuration of C-4. (This atom becomes C-2 in the furan nomenclature.)

Conversely, another dioxolenium ion (**IX**) may emerge from the opening of the epoxide ring of **I,** whereby inversion of configuration now occurs at C-4 (see Scheme 48.3). This yields a (*R*)(*R*)(*threo*) structure instead of the (*S*)(*S*) (also *threo*) structure **III.** This new intermediate (**IX**) may also be depicted in its three-dimensional representation as in **X.** It clearly is open to evolution in the same sense indicated before, to give ortho ester **XI**—or conceivably the unobserved oxetane **XII**. Again **XI** represents a division in the pathway that leads to tetrahydrofurans: Path **E** yields another wrong oxacyclopentane **Vb** via **XIII,** whereas path **F** affords a second correct *trans*-substituted **II,** that is, **VIIIb,** by way of configurational inversion of C-4.

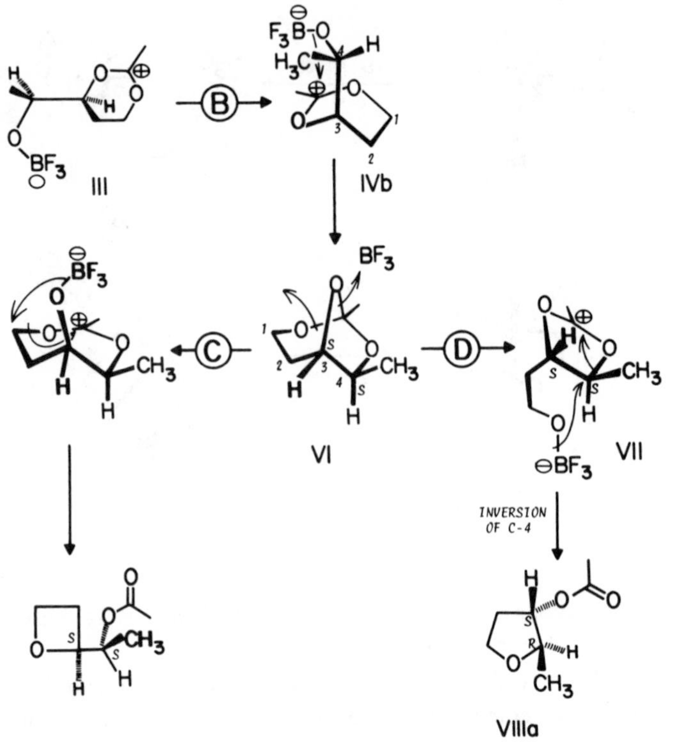

SCHEME 48.2

Amusingly enough, we have managed to devise not one but four more or less acceptable routes, arriving at two hypothetical (**Va**, **Vb**) and two real (**VIIIa**, **VIIIb**) structures by quite dissimilar approaches. Mechanistically speaking, this predicament is unacceptable and demands further experimental clarification. Fortunately, labeling experiments have been performed using ^{18}O tracers specifically at the acetate carbonyl of **I.**[5] After hydrolysis of **II** the label was found in the alcohol by-product. The only route that accounts for this observation is the sequence **B**, **D** of Scheme 48.2. As a consequence, were compound **I** really optically active (it was not in the actual experiment), only the optically active **VIIa** should have been obtained. Since asymmetric syntheses are an important subject of research nowadays, the real experiment should be run.

Now, the apparently innocuous insertion of a methylene in the acetate bearing chain that was mentioned at the beginning of this discussion, namely, com-

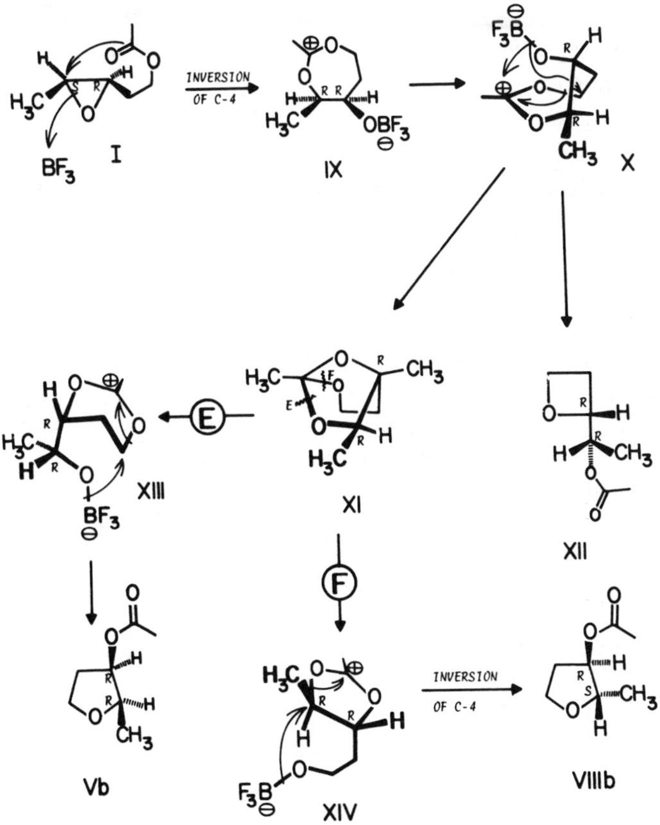

SCHEME 48.3

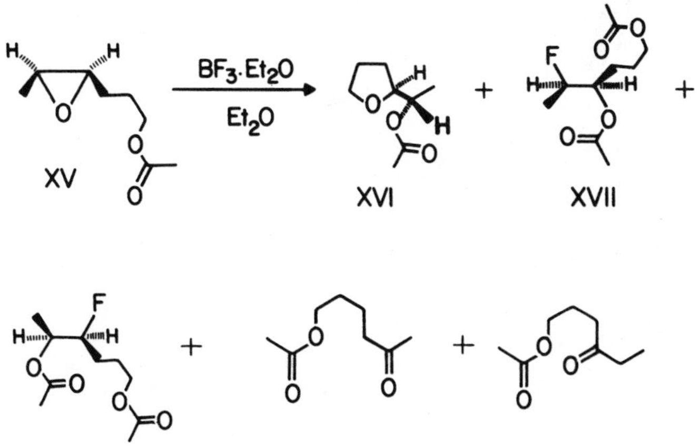

SCHEME 48.4

pound **XV,** yields a whole host of products (see Scheme 48.4) on exposure to boron trifluoride etherate, instead of the anticipated tetrahydropyranyl derivative that would be expected from the extrapolation of the mechanism just described. The interested reader is invited to solve this stimulating problem. (Yes, F in structures **XVII** and **XVIII** means fluorine, it is not a printing error.)

REFERENCES

2. These species have been observed by nmr spectroscopy; see G. A. Olah and J. M. Bollinger, *J. Amer. Chem. Soc.*, **89,** 4774 (1967).
3. A Parsec is 206,265 times the distance from the planet earth to the sun, or 3.26 light-years, or 19.2 trillion miles. It is used in astronomy as a standard unit for measurements of stellar distances.
4. For the rationalization of a related transformation see A. F. Thomas and W. Pawlak, *Helv. Chim. Acta*, **54,** 195 (1971).
5. This experiment was actually performed with a sample of *trans* epoxide instead of the *cis* derivative.[1] Yet the results are equally valid, since the oxygen label has no bearing on the stereochemical course of the reaction.

PROBLEM 49

I

$\xrightarrow{1,2}$

II + III

1 = $HO(CH_2)_2OH$, THF, >=O, NBS, -15°, 3 min
2 = $NaHSO_4$, H_2O, 55°, 30 min

---

1. C. Fehr, G. Ohloff, and G. Buchi, *Helv. Chim. Acta*, **62,** 2655 (1979).

PROBLEM 49
Oxidative versus Nonoxidative Conversion of a Macrocyclic Tosyl Hydrazone

This interesting reaction provides an excellent opportunity to revise once again the all-important concept of relative oxidation levels. Comparison with the process portrayed in Problem 8 is convenient now because both reactions describe closely related phenomena, which, however, differ markedly in their results.

The two sequences have been drawn in Scheme 49.1 in a rather prejudiced manner in order to emphasize the most striking similarities (parts **A** and **B**). Also, the transformation of **I** has been artificially portrayed in such a way that no reagents other than alcohol appear on the scheme (part **B**). Notice first that both substrates are identical α,β-unsaturated tosyl hydrazones, since **IV** is the cyclic equivalent of compound **I** in Problem 8. Still, one leads to an alkene while the other yields a more highly oxidized product **XV**.

Furthermore, it is of interest to note that the overall process that leads to intermediates **VI** and **VIII** here are basically equivalent in the sense that both are the consequence of the attack by a nucleophilic entity at the imine carbon atom. However, while the introduction of alkoxy at this carbon is the equivalent of a S_N2' substitution, the introduction of hydride requires the more circuitous and conceptually different path of reductive addition followed by nonoxidative elimination. In consequence, the latter implies reduction, the former does not.

If from this point onwards the parallelism of the two pathways were to continue, the corresponding 1,5-hydrogen shift experienced by **VIII** (route **A**) would yield enol ether **IX,** which would evolve into the saturated ketone **X.** Conversely, the extrusion of molecular nitrogen as indicated in route **B** would furnish an α,β-unsaturated ketone (**XI**). This fact embodies the production of one equivalent of hydride, obviously an oxidative step that is reflected in the higher oxidation level of **XI** as compared with **X.**

Before taking care of this hydride further, let us repeat the two processes just described using structure **XII** as an imaginary starting material. This compound would be simply a tautomer of **VIII.** While the 1,5-hydrogen transfer (route **C**) should yield alkoxyacetylene **XIII,** the extrusion of nitrogen and hydride (route **D**) would afford compound **XV,** which features the same oxidation level as the target structure **II.**

Hydride ions are not easily expelled from organic molecules. In the present case the situation would be more easily manageable if this hydrogen were replaced by a good leaving group. This is easily achieved if one waves the

SCHEME 49.1A

SCHEME 49.1B

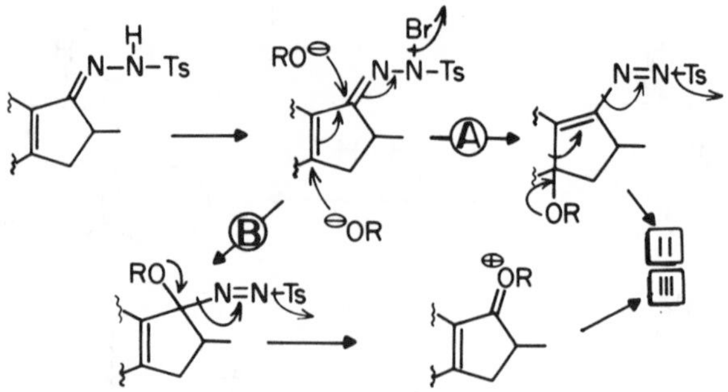

SCHEME 49.2

self-imposed restriction of not adding reagents other than alcohol to the reacting mixture. Lost among an assortment of solvents there is *N*-bromosuccinimide (NBS) available for introducing a bromine atom precisely where the hydride, so central in our previous discussion, is located. This process would be analogous to the well-known allylic bromination for which NBS is so successful.[2] In fact, this *N*-bromination was devised in an earlier work[3] as a practical means to convert tosyl hydrazones to ketones. The more probable mechanism that mediates between **I** and ketones **II** and **III** would be that portrayed on Scheme 49.2.

The description of a few instructive pieces of information are in order. If the oxidation level of tosyl hydrazone **I** is raised one step, say by the conversion of the C=C bond into an oxirane, and if the reader follows a route analogous to pathway **C** with this new hypothetical starting compound, it will become apparent that a keto aldehyde similar to **XV** arises, though not necessarily through route **D**. What has been done in essence is to transfer the oxidation level required to supersede the hydride extrusion step to the carbon backbone itself. This makes the use of NBS or any other electrophile unnecessary. What the reader runs into in doing this is one of the most pervading reactions of recent years called the Eschenmoser cleavage of α,β-epoxy ketones.[4] In fact, the reaction that constitutes the present problem was developed as an alternative of the inveterate Eschenmoser synthesis when the epoxidation of the enone is impeded by steric overcrowding.[5]

REFERENCES

2. For a review see I. Horner and E. H. Winkelmann, *N-Bromosuccinimide, Its Properties and Reactions. The Course of Substitution*, vol. III, W. Foerst, Ed., Academic, New York, 1964 p. 151.

3. G. Rosini, *J. Org. Chem.*, **39,** 3504 (1974).
4. For leading references see A. Eschenmoser et al., *Helv. Chim. Acta*, **50,** 2101 (1967); **51,** 1461 (1968); **53,** 1479 (1970); **54,** 2896 (1971); **55,** 1276 (1972). For more recent developments see, among others, E. J. Corey and H. S. Sachdev, *J. Org. Chem.*, **40,** 579 (1975); P. L. Fuchs, *J. Org. Chem.*, **41,** 2935 (1976).
5. For those who feel puzzled by the introduction of an acetylene fragment in a cyclic structure, they are invited to meet *nonbornyne*. See P. G. Gassman and J. J. Valcho, *J. Am. Chem. Soc.*, **97,** 4768 (1975).

PROBLEM 50

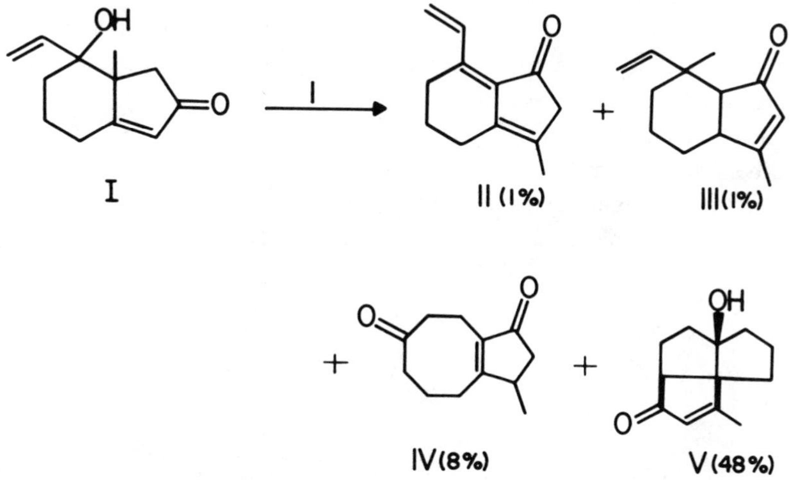

I

II (1%)

III (1%)

IV (8%)

V (48%)

I = CH_3OH, H_2O, 0.13M KOH, Δ, 9 min

1. W. G. Dauben and D. J. Hart, *J. Org. Chem.*, **42,** 3787 (1977).

PROBLEM 50
Fragmentation of an Enone during an Aldol Condensation

This reaction appears to be of the kind where molecules are torn apart and new compounds are reconstructed arbitrarily from the remaining molecular bits. The only feature that the four products and the starting enone have in common is the cyclopentanone fragment. However, the substitution pattern and the position of the double bond of this carbocycle in compound **I** does not find any correspondence in the products. It is also meaningful that the quaternary carbon that bears the angular methyl ends up as a tertiary center in all products **II–V.** Evidently, the fragmentation of the C–C bond between the carbinol and its vicinal quaternary center is in order. In that case, one has a means of liberating a six-carbon side chain with a polar function at the end, from which various C–C bond forming reactions are imaginable.

Now, any fragmentation reaction requires an electron sink where the released electrons can go. Compound **I** contains one such drain of negative charge

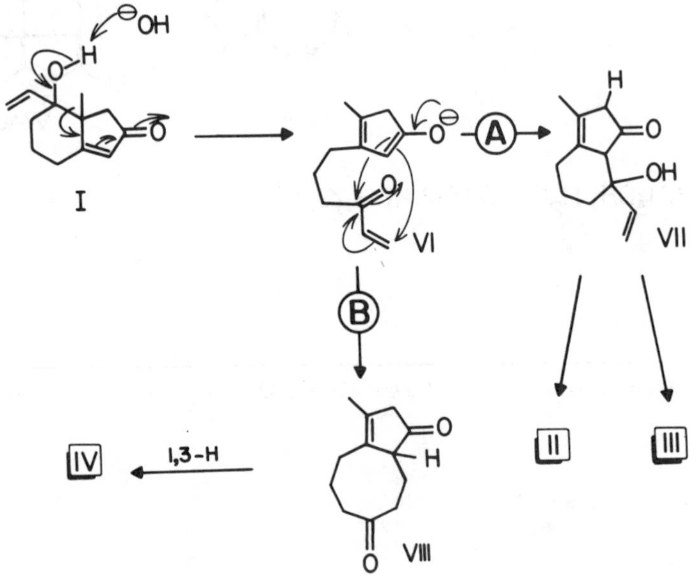

SCHEME 50.1

in the enone moiety. Once this is realized, it should not be difficult to configure a fragmentation of the ring to the key enone **VI** by way of a vinylogous retro-aldol condensation.

The relative positions of the carbanion—or enolate—and the carbonyl group created in the side chain by this fragmentation, on one hand, and the preservation of the monosubstituted ethylene fragment on the other, are strongly suggestive of the occurrence of an aldol condensation. This would terminate inevitably in products **II** and **III** (route **A** of Scheme 50.1) and the eight-membered ring of **IV** would result from the alternative 1,4 addition to this enone, followed by migration of the double bond via 1,3-hydrogen shift (route **B**).

If we now concentrate our attention on cyclopentenone **V** and compare this product with **IV** it will be obvious that another intramolecular aldol condensation between the cyclooctanone ketone and the angular carbon β to the cyclopentenone keto function will suffice to convert **IV** into **V,** with the proper stereochemistry. The required nucleophilic character of the carbon involved there would be derived from the base assisted proton abstraction in a kinetic fashion

IV

IXa

VIII

IXb

V

SCHEME 50.2

indicated in **VIII** (see Scheme 50.2). In addition, the approach of the two interacting units would be facilitated by the well-known flexibility of the cyclooctane ring (see **IXb**).[2,3]

REFERENCES

2. For a review see A. C. Cope, M. M. Matin, and M. A. McKervey, *Q. Rev. Chem. Soc.*, **20,** 110 (1966).
3. For a similar rearangement, see P. V. Ramani, J. P. John, K. V. Narayanan, and S. Swaminathan, *J. Chem. Soc. Perkin 1*, 1516 (1972) and references cited therein.

PROBLEM 51

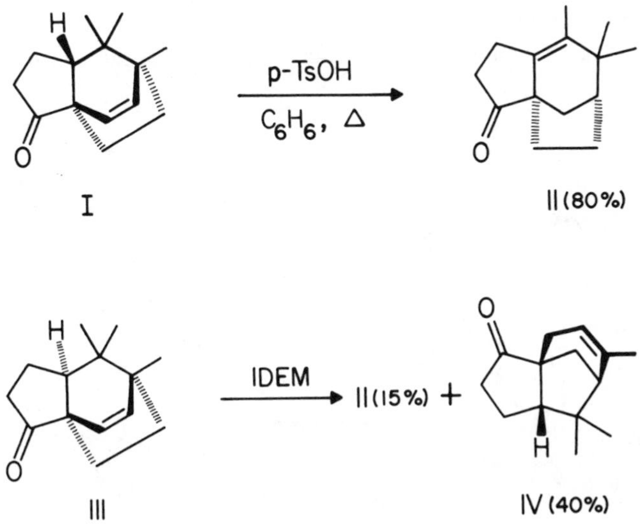

p-TsOH

C_6H_6, Δ

I

II (80%)

IDEM

II (15%) +

III

IV (40%)

1. G. Buchi, A. Hauser, and J. Limacher, *J. Org. Chem.*, **42,** 3323 (1977).

PROBLEM 51
A Potentially Complex Wagner–Meerwein Rearrangement Made Simple

Carbon to carbon σ-bond migration processes, which are visibly involved in the transformations of **I** and **III,** usually offer savory mechanistic problems whose main difficulty—most people think—resides in the drawing of intermediate structures. These are often complicated by the migration of bonds and atoms, and require the foresight to conceive convoluted spatial transformations. This sort of problem, however, features another major hardship: If one makes generous use of one's imagination, a whole series of carbocationic intermediates

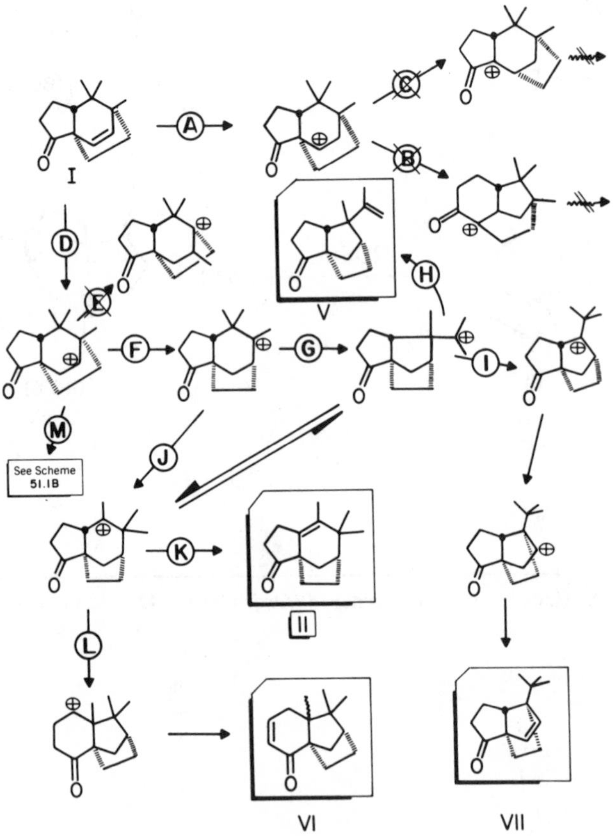

See Scheme 51.1B

SCHEME 51.1A

that split to yield more and more carbenium ions will be conceived. This is translated into a vast number of possible products. The real problem then becomes not how to account for observed products, but how to develop valid criteria to be able to decide which of the many possible products and reaction routes can be expected from the acid treatment of a rearrangement-prone olefin.

The reactions of this problem are good examples of this occurrence. Compounds **I** and **III,** paradoxically, display the potential for yielding a host of products while they behave cleanly in the actual experiment. It is of additional interest to note the control over the stereochemistry of the zero bridge in **I** and **III.**

Any further comment will be withheld to allow the readers to show their

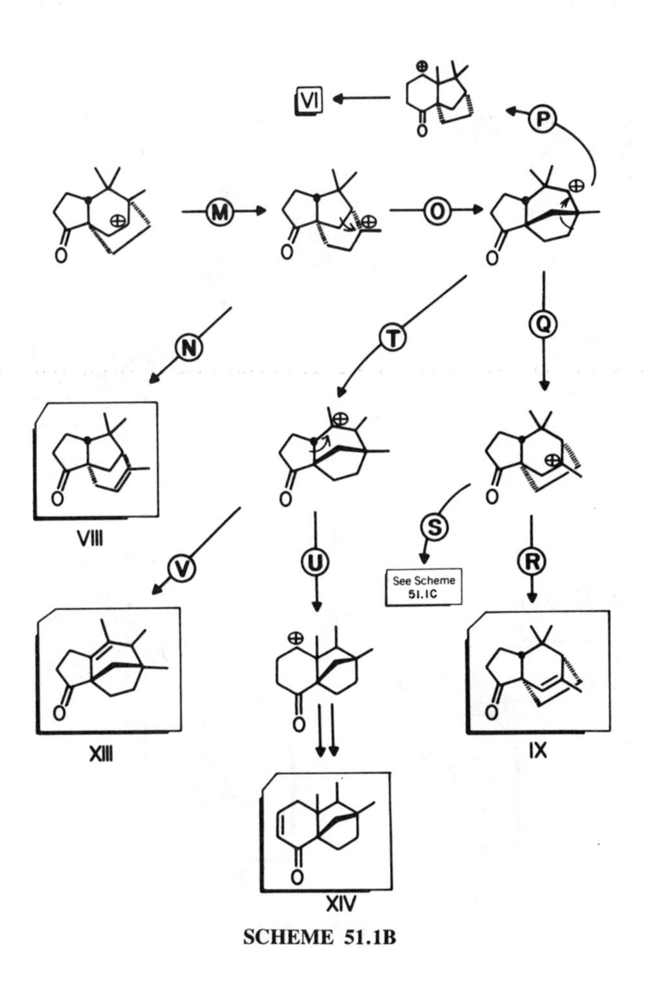

SCHEME 51.1B

ability as C–C bond migrators. Then we shall compare the reader's results with our mechanistic tree (see Scheme 51.1).[2]

At least 11 compounds (and there are probably more) can be theoretically conceived during the acid treatment of **I** by way of rearrangement pathway **D** alone. Route **A** may not be all that relevant since it leads to unstable cations vicinal to carbonyl groups. Also, not all the intermediates contrived in the previous scheme are stable entities, owing chiefly to those stereochemical restrictions imposed by the bicyclic system. For example, the *trans*-ring fusion of the cyclopentanone ring and its adjacent bicyclic structures have considerable ring strain. This becomes visible in the three-dimensional representation of structure **VIII,** that is, **XVI** that would be the epimer of the observed product (**IV**).

This being the case, all those routes that lead to compounds contaning one

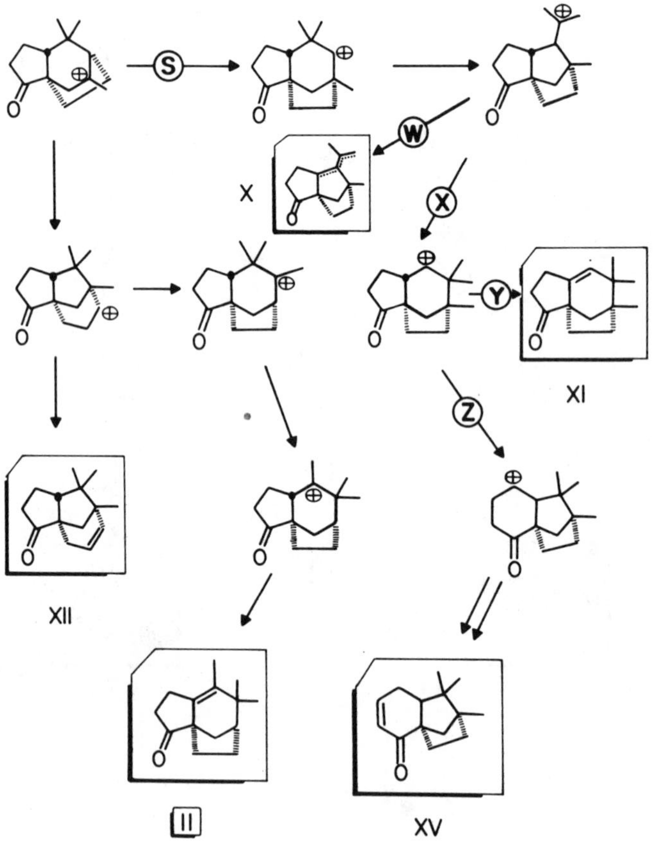

SCHEME 51.1C

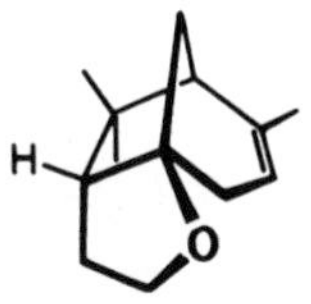

XVI (VIII) XVII (IV)

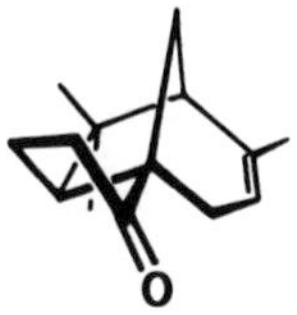

such *trans*-ring fusion, namely, **V, VII, VIII, X,** and **XII,** although still feasible, would be much less competitive. In addition, other products require the conversion of tertiary to secondary carbenium ions. Despite the chemical lawfulness of one such operation, the energy requirements are higher than for other routes and consequently these paths may be disregarded as well. Therefore, one may cross out anything produced after steps **L** and **O.** The unique survivor of this erasing process is, fortunately enough, observed product **II.**

The situation becomes considerably more complex for compound **III** because the *cis*-ring fusion portrayed in **XVII** confers enough stability for this product to be isolated. Considering all other limitations mentioned previously, at least structure **V** should have the right to exist. However, the additional ring strain associated with the tricyclo [4.3.1.0] system is likely to militate against its formation from **III** under the recorded conditions.[3]

REFERENCES

2. We have imposed the restriction of using only secondary or tertiary carbenium ions and have avoided bridgehead carbocations, only because of their higher instability. This comment is in order in view of recent evidence of their existence. See, for example, Problem 44.
3. For an excellent formal treatment of the Wagner–Meerwein rearrangements of camphor in sulfuric acid, which makes extensive use of ^{14}C labeling, see O. R. Rodig and R. J. Sysko, *J. Am. Chem. Soc.*, **94,** 6475 (1972).

PROBLEM 52

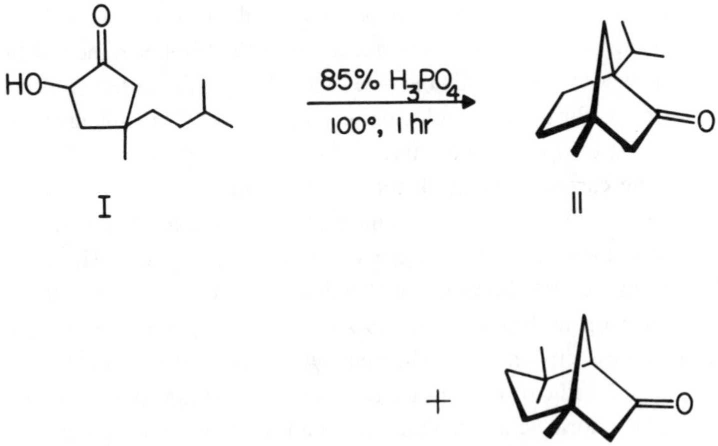

$$\text{I} \xrightarrow[100°,\ 1\ \text{hr}]{85\%\ H_3PO_4} \text{II} + \text{III}$$

I

II

III

---

1. W. C. Agosta and S. Wolff, *J. Org. Chem.*, **41**, 2605 (1976).

PROBLEM 52
Strange Involvement of Nonfunctionalized Isopentyl Side Chain

Hot, concentrated mineral acids are splendid reaction media for obtaining surprising results. As we shall now see, this case goes even beyond the anticipated astonishment.

Compounds **II** and **III** are isomeric, yet different enough to suggest that they require divergent mechanisms for their formation from **I.** Counting C–C bonds in the products reveals that these compounds have one more C–C bond than the starting material. This is in accord with the disappearance of the carbinol from **I,** whose carbon fills the role of an electrophile rather well in acidic media. However, for the intramolecular construction of C–C linkages, one also needs a function of opposed polarity, in this case a nucleophile. The only one available is the carbon α to the ketone in its enolic form. Yet, this situation cannot account for the strange involvement of the isopentyl side chain that ought to take place in the transition from acyloin **I** to ketones **II** and **III.**

The structure of **III** shows that a C–H bond, presumably that of the tertiary carbon of the isopentyl substituent, is somehow being replaced with a C–C linkage, a process customarily achieved by means of free radical couplings. However, such a radical, as well as any sort of carbanion, would be unlikely in aqueous phosphoric acid. There seem to be few alternatives left at this point.

There would be a way out if only compound **I** were able to deliver a carbocation at the tertiary position of the isopentyl chain. In turn, this would only require the hydride transfer from this position to a very low electron density center in the neighborhood of this pivotal carbon. Acyloin **I** can fulfill this purpose when it becomes associated with phosphoric acid. On one hand, the α carbon of the enol in **IV** must first reverse its normally negative polarization to allow for the introduction of the hydride at this point. Such an odd role may be better understood if this enol is looked upon as an allyl cation **V** (see Scheme 52.1).

On the other hand, cation **VI** would also serve as a receptor of the hydride ion from the side chain, whose folding over the carbocycle would place the two reactive sites at very close range. The local reductive step that this hydride transfer implies would be compensated for by the ensuing elimination of water in **VII** that regenerates the keto group. The formation of the cationic center then secures the construction of **III** from this key intermediate.

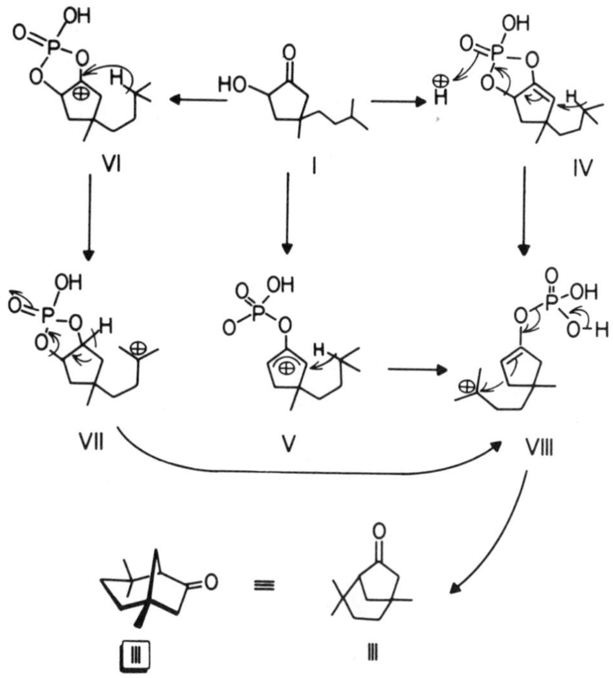

SCHEME 52.1

The path leading to **II** follows a more circuitous route that may not become directly apparent unless the central cation **VIII** is assumed to participate. One has to figure out first, however, how a methylene of the side chain is activated properly for C–C bonding, since it is now this carbon that is involved in the production of **II.** (Before continuing, the reader must be convinced of this.) Its activation may be accomplished by rather standard means, namely, E1 elimination that would yield olefin **IX** followed by anti-Markovnikov addition of a proton. The secondary carbocation thus formed then would be trapped by the enol function in **X,** configured now in the opposite direction to that in **IV** (see Scheme 52.2).

The problem posed by the energy costly anti-Markovnikov addition may be circumvented by an intellectually more rewarding scheme. Knowing that bicyclo[2.2.1]heptane systems are converted to other bicyclo[2.2.1]hepanes in acidic media with relative ease on one hand, and realizing that the carbonyl and carbinol carbons in **I** are equivalent as far as linking places for the side chain on the other, one can imagine that bicyclic compound **II** may result from **XI**

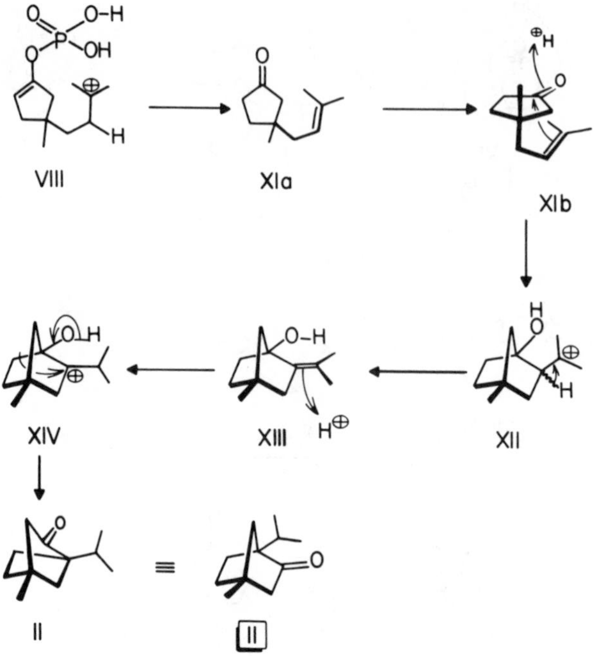

SCHEME 52.2

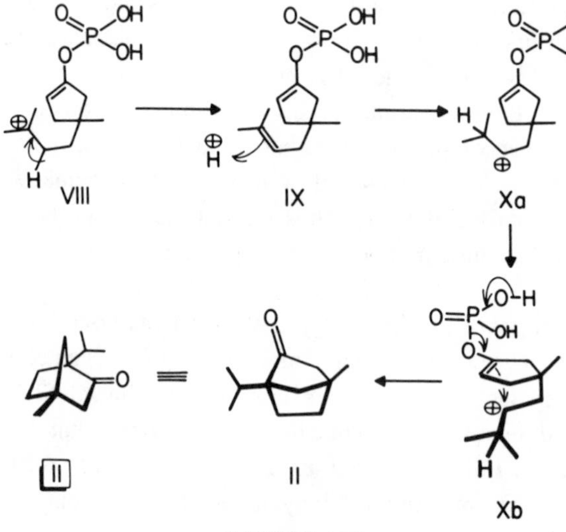

SCHEME 52.3

by way of a Prins reaction.[2] Here the C=C bond is used as the source of electron density for the construction of the new C–C linkage, thus producing a more favorable carbenium ion. Again, elimination and a Markovnikov addition would place the carbenium ion on the ring at a comfortable tertiary position, precisely where it is needed for the ensuing pinacol rearrangement that leads to **II** (see Scheme 52.3).

REFERENCES

2. For a review see D. R. Adams and S. P. Bhatnagar, *Synthesis*, 661 (1977).

PROBLEM 53

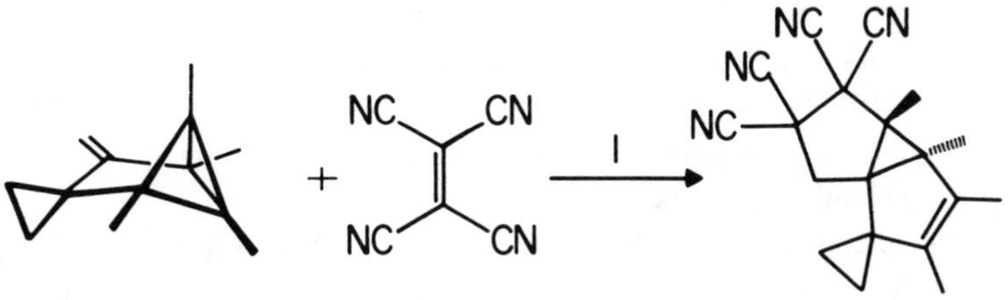

I + TCNE $\xrightarrow{\text{I}}$ II

I = $CHCl_3$, 20°, 15 min

1. R. F. Heldeweg and H. Hogeveen, *J. Org. Chem.*, **43,** 1916 (1978).

PROBLEM 53
First Encounter with Woodward–Hoffmann Symmetry Rules: How These Help in Choosing the Correct Answer

Occasionally one comes across transformations whose visualization in terms of formal steps that follow chemically possible transformations cannot be figured out readily, simply because too many atoms and bonds appear to be involved. It is necessary first to detect which bonds and atoms undergo change, and then to add some degree of organization to these steps. In order to accomplish this it is advisable to draw substrates and products in the way in which they resemble each other the most. If this is still not sufficient, then we should for once think of bonds and atoms as toothpicks of a ball and stick molecular model and decide how to move them about in order to correlate the two structures. Finally, a serious attempt to adapt these redistributions of sticks to chemically allowed operations should be the next step.

We were compelled to use this procedure in seeking a solution to this problem. In addition, fragmentation analysis (Scheme 53.1) provided useful information such as the obvious identification of tetracyanoethylene (TCNE) and vinyl cyclopropane fragments in the product. It also indicates that TCNE becomes bonded to the terminal olefin by a process other than [2+2] cycloaddi-

NC CN NC CN NC CN NC NC

SCHEME 53.1

tion. As a consequence, the second C–C bond between TCNE and the substrate must be derived from a still unclear participation of the bicyclobutane portion of the molecule.

In this case, two-dimensional representations of starting material and product are more convenient than the spatial drawings for correlation analysis. Once this is done one can see that there are not one but at least four ways to move about the little sticks representing C–C bonds to convert **I** into **II** (see Scheme 53.2).

To translate this schematic reasoning into sound chemistry in such a complicated case one cannot naively place pluses and minuses here and there to indicate positive and negative charge development. System **I** demands careful analysis in terms of the Woodward–Hoffmann symmetry rules[2] since this is a thermally induced reaction.

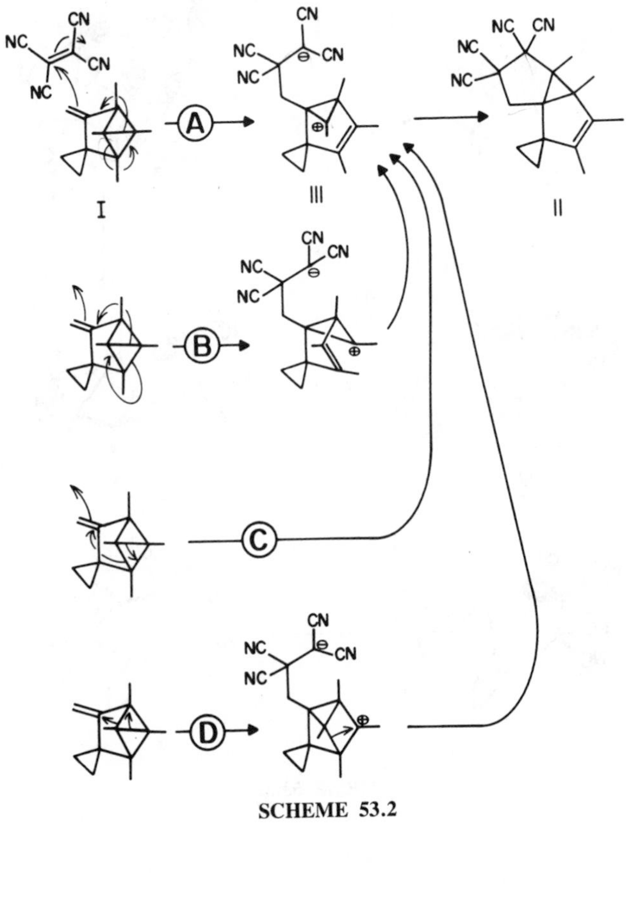

SCHEME 53.2

Our first approach could be to attempt to mitigate the strain associated with the bicyclobutane ring system by disconnecting its central bond. This is represented by routes **A** and **B** of Scheme 53.2. These are based in a 1,3-sigmatropic shift of this central bond with disrotatory ring opening (rotation in opposite directions). This is allowed by the Woodward–Hoffmann selection rules in a thermal process as long as it takes place suprafacially with inversion of configuration of the migrating carbon (denoted in **V** with an asterisk) (see Scheme 53.3). This satisfies the severe stereochemical restrictions of compound **I.**

This situation implies that **I** reacts as a neutral species. That is, its interaction with TCNE occurs simultaneously with the 1,3-shift as portrayed in **VI.** A cation **VII** results from this step, which evolves towards the final product by

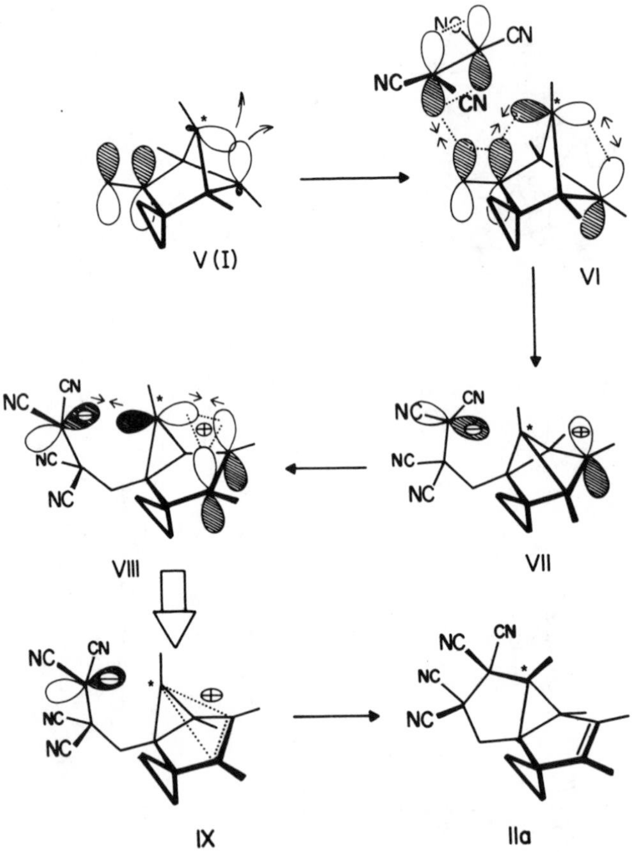

SCHEME 53.3

way of a nonclassical carbocation (portrayed as either **VIII** or **IX**) constituted by three atomic π orbitals and two electrons.

This operation is in essence an incomplete, symmetry allowed suprafacial 1,2-sigmatropic shift, where the low electron density that develops progressively at the *tip* of the cyclopropane ring is compensated by the favorably located dicyanocarbanion. This prevents the transformation of **IX** into the hypothetical cation **X,** a potential precursor of divergent reactions.

The migration of the central C–C bond of the bicyclobutane portion of **I** may also take place in the opposite direction (route **B** of Scheme 53.2), in view of its symmetrical disposition with respect to the double bond. Thus, structure **V** that may also be represented as **XI** would proceed towards the target compound by way of a bimolecular attack on TCNE and through the same sort of intermediate steps described previously (see Scheme 53.4). The final product **IIb,** however, would be enantiomeric with respect to the end product in Scheme 53.3 (namely, **IIa**).

Although a third mode of symmetry allowed 1,2-sigmatropic shift (route **C** of Scheme 53.2) is still possible, probably it would not be as favorable as the former two, chiefly because, while the former two occur within a vinylcyclopropane framework, route **C** entails a homovinyl cyclopropane ring system, whose chemistry is quite different. The reaction of **I** with TCNE still may take a completely different course if the intervention of ionic species prior to the required carbon migrations mentioned earlier is allowed. This idea would be underscored by the strong stabilization of the resulting dipole that would feature two efficient electron attracting nitriles at the negative end, and two moderately efficient electron donating units (cyclopropanes) in the most appropriate bisected conformation for orbital overlap (maximum stabilization power) at the positive end.

In this case only two electrons and a highly polar transition state would be involved in the ensuing bond migrations. The Woodward–Hoffmann selection

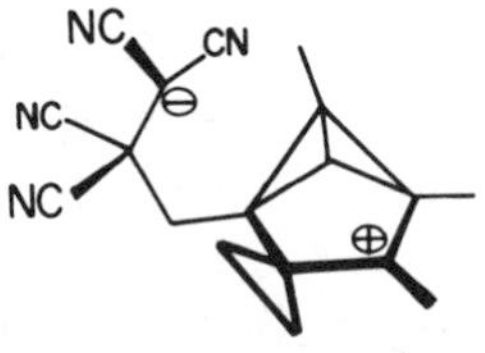

X

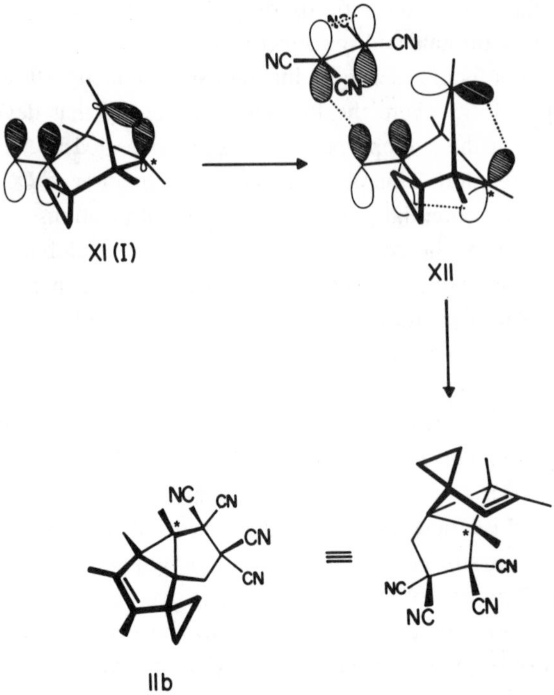

XI (I)

XII

IIb

SCHEME 53.4

rules have the 1,2-sigmatropic alkyl shift as a more favorable process than the 1,3-counterpart. This paves the way for route **D** suggested in Scheme 53.2. A sequence of two suprafacial 1,2-shifts with retention of configuration represented by **XIV** and **XVI** (see Scheme 53.5) converges toward the familiar intermediate **VII,** which in turn evolves into **IIa.**

The readers may convince themselves that enantiomeric product **IIb** emerges from the allowed 1,2-alkyl shift of the other cyclopropyl C–C bond of **I.**

The available experimental information does not warrant any definitive distinction between these four possibilities. However, some indirect evidence allows for some choices. Regardless of the 66 kcal/mol of strain embodied in the bicyclobutane ring, it has been shown to be a relatively stable system in the gas phase or in hydrocarbon solvents.[3] Nevertheless, when bicyclobutane is heated at 200°C two of the peripheral C–C bonds are broken, while the central bond remains intact during the purely thermal, symmetry controlled reactions. This relative stability has been associated with the predominantly π character of this

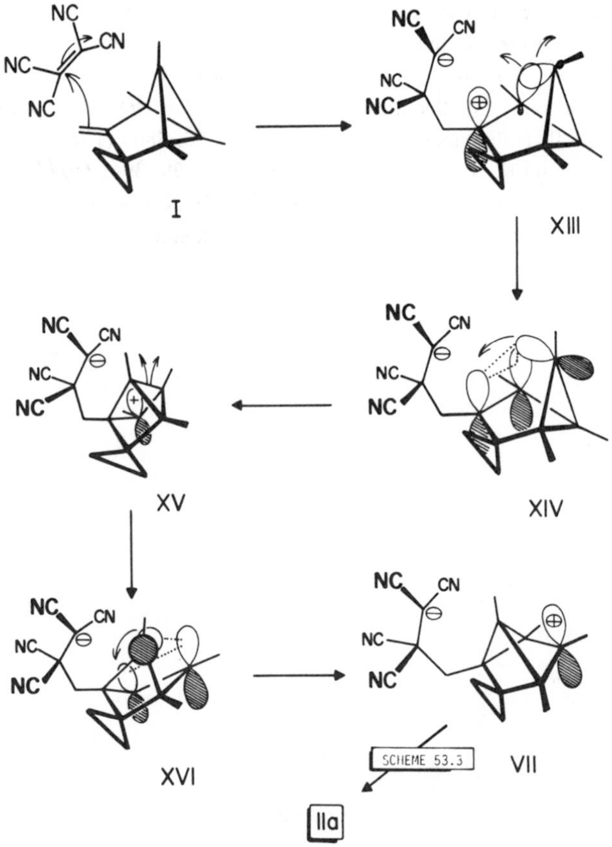

SCHEME 53.5

central bond,[4] a very peculiar feature that has served as a basis for the construction of some transition metal complexes of bicyclobutane.[5]

Along these lines, route **D** (see Scheme 53.5) is the only pathway in which the central bond of the bicyclobutane fragment is broken after one of the peripheral C–C bonds is disconnected. This forces it to lose most of its π character and becomes easier to separate. Route **D** appears, therefore as the more favorable.

REFERENCES

2. R. B. Woodward and R. Hoffmann, *The Conservation of Orbital Symmetry*, Verlag-Chemie/Academic, Weinheim, 1970. Several other excellent textbooks deal with this

central subject. See, for example, T. L. Gilchrist and R. C. Storr, *Organic Reactions and Orbital Symmetry*, Cambridge University Press, Cambridge, 1979.

3. For a review see K. B. Wiberg, *Adv. Alicyclic Chem.*, **2,** 185 (1968).
4. M. D. Newton and J. M. Schulman, *J. Am. Chem. Soc.*, **94,** 767 (1972); M. Pomerantz and D. F. Hillebrandt, *J. Am. Chem. Soc.*, **95,** 5809 (1973); K. B. Wiberg, G. B. Ellison, and K. S. Peters, *J. Am. Chem. Soc.*, **99,** 3941 (1977).
5. H. Takaya, T. Suzuki, Y. Kumagai, M. Hosoya, H. Kawauchi, and R. Noyori, *J. Org. Chem.*, **46,** 2854 (1981) and references cited therein.

PROBLEM 54

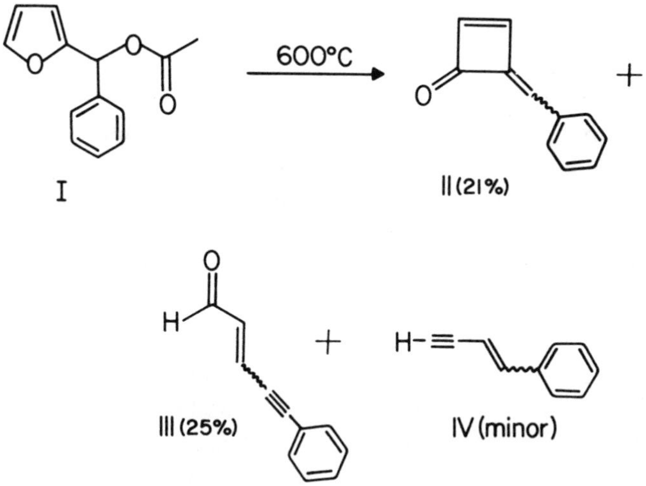

600°C

I

II (21%)

III (25%)

IV (minor)

1. W. S. Trahanowsky and D. L. Alexander, *J. Am. Chem. Soc.*, **101,** 142 (1979); W. S. Trahanowsky and M. -G. Park, *J. Org. Chem.*, **39,** 1448 (1974).

PROBLEM 54
Second Encounter with Woodward–Hoffmann Symmetry Rules: Double Migration of the Same Acetoxy Group

For most organic compounds the molecular collisions that occur at 600°C have the effect of a sledgehammer on a peanut. For this reason, pyrolytic transformations offer ample opportunity to advance the most adventurous mechanistic speculations. The reaction presented here is a good example of this. It demands imagination not only to account for observed products and to find a reasonable explanation for the survival of the highly unstable methylene cyclobutenone (**II**) at such a high temperature,[2] but also to adapt the scheme to the Woodward–Hoffmann symmetry rules.

At the outset, it might be useful to recognize that products **II** and **III** contain all five carbons of the furfurylidene backbone of **I** along with the phenyl substituent. The only missing component is a molecule of acetic acid, which implies that, in addition to the acetate substituent, we have to pick up a proton from a still undefined place on the molecular core.

Second, comparison of **II** and **III** suggests that the latter may be converted into the former if just the acetylene and aldehydo groups were to form a C–C bond while, at the same time, the aldehydic proton was transferred to the benzylic position. If compound **III** would attempt to do this by trapping the proton in question with the triple bond, the ensuing cyclization would demand the use of the carbonyl carbon as a nucleophile, something that is highly unlikely.

This difficulty may be overcome by the conversion of **III** to a more manageable intermediate. The structure of this mediator may be figured out by working backwards the route that leads to enone **II** (see Scheme 54.1). Of the two possible modes of retrocyclization of **II**, the one that keeps all the carbons in one continuous chain, represented by path **B** of Scheme 54.2, results in cum-

SCHEME 54.1

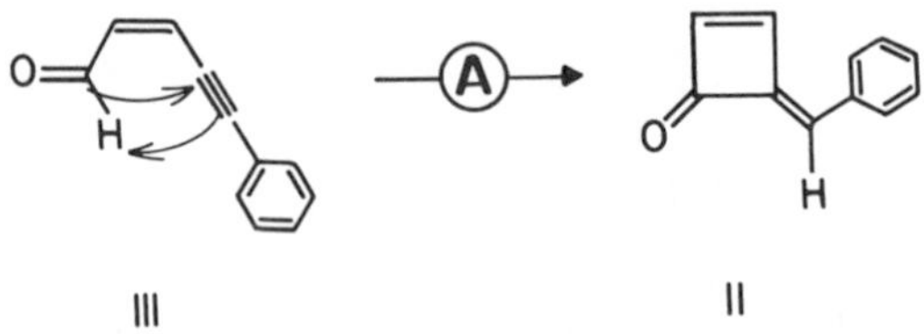

O H III —(A)→ O H II

SCHEME 54.1

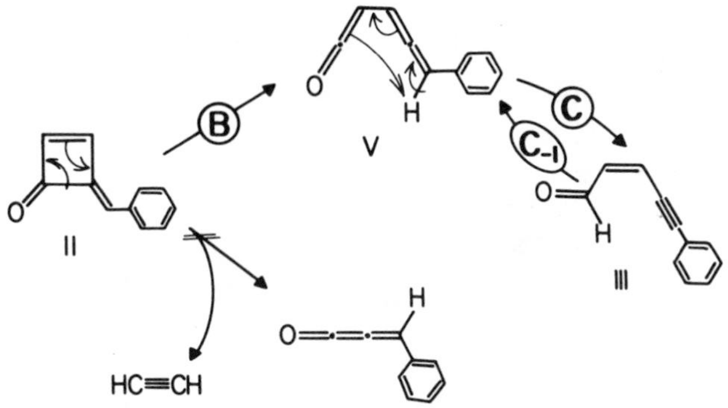

SCHEME 54.2

mulenoketone **V.** In turn, this intermediate would be consequential to the now feasible 1,5-hydrogen shift of the terminal aldehydic proton to the benzylic carbon of **III** as shown in route **C.**

All this means that the mechanism will be elucidated once a reasonable channel for the conversion of **I** into either **III** or **V** is found. One might start this search by realizing that furan acetate **I** contains all the components needed to construct the acetylene moiety of **III.** This becomes clear if one reorganizes electrons as shown in the sequence **D–E** (see Scheme 54.3).

Translation of this conception in terms of the actual starting material would find a reasonable beginning at the cyclic oxygen-assisted elimination of acetate to give **XI,** and proton elimination (see Scheme 54.4).

This clever solution would be perfect were it not for the fact that, having no aid from solvent molecules, organic intermediates are remarkably reluctant to yield ionic species in the gas phase, no matter what tight-ion pair or whatever other charge buildup preventing molecular artifice we use. In other words, in

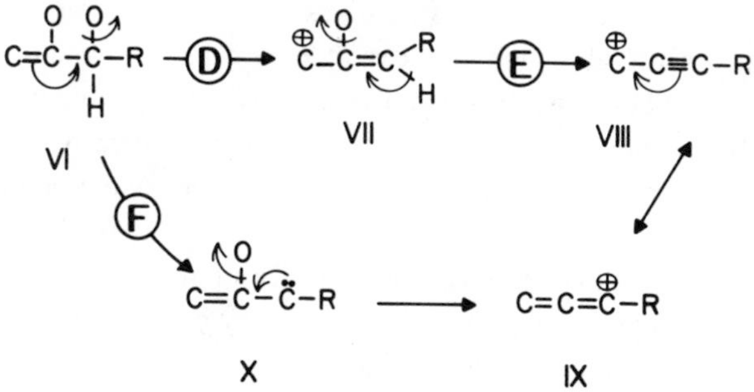

SCHEME 54.3

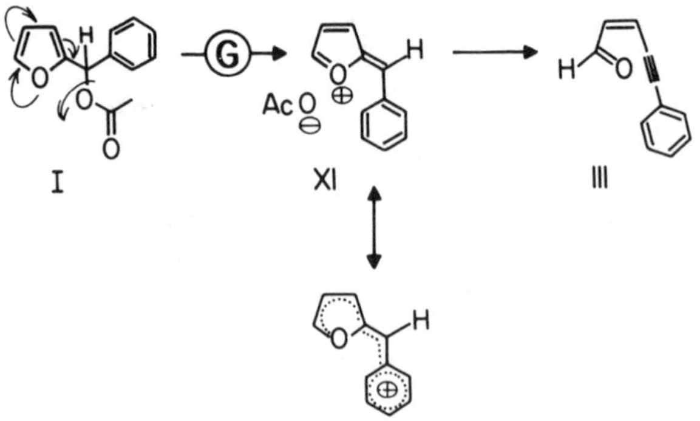

SCHEME 54.4

the gas phase a mechanism with a high degree of concert with little charge polarization is desirable.[3]

With this restriction in mind, other solutions of the same central idea and without the participation of free radical species are conceivable. For instance, 1,4 elimination of acetic acid shown in route **H** would yield ketene **V** directly, a contention that finds support in the favorable 1,5 elimination portrayed in **XII** (see Scheme 54.5).[2] Analogously, a benzylic carbene precursor would also be in a position to give an acetylene if route **F** of Scheme 54.3 is handled in such a way as to prevent charge development.[4] In fact, carbene **XIV** that would result from the 1,1 elimination of acetic acid from **I**,[5] has been shown to give aldehyde **III** when furyl–phenyl diazomethane (**XV**) (an efficient carbene generator) was used as precursor.[6]

There are still two more propositions that can find room in the present scheme. Both are based on the treatment of the would be aldehydic carbon at this oxidation level in the form of a mixed acetal. This situation would be achieved by somehow placing the acetoxy group at C-5 of the furan nucleus. One plausible means of doing this involves a 3,5-sigmatropic rearrangement. This particular shift of which few examples are known would be thermally allowed by the Woodward–Hoffmann rules only as a suprafacial–antarafacial migration. In the present case, the geometry of the corresponding transition state is nearly impossible to achieve. This casts serious doubt on the occurrence of this route.

Furthermore, a rather bold but chemically more feasible mechanism can account for the proposed migration of the acetoxy group. This is double, symmetry allowed, suprafacial–suprafacial, Claisen-type, 3,3-sigmatropic shift.

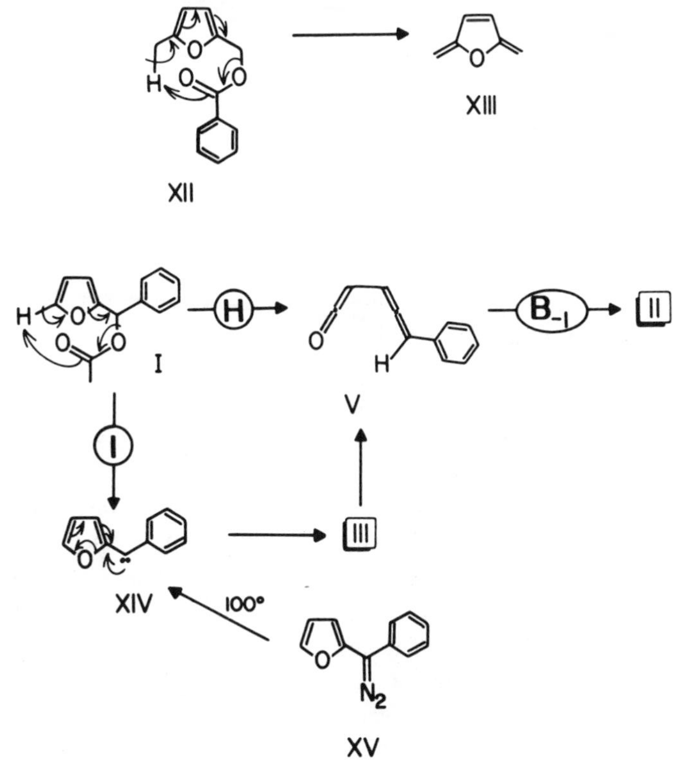

SCHEME 54.5

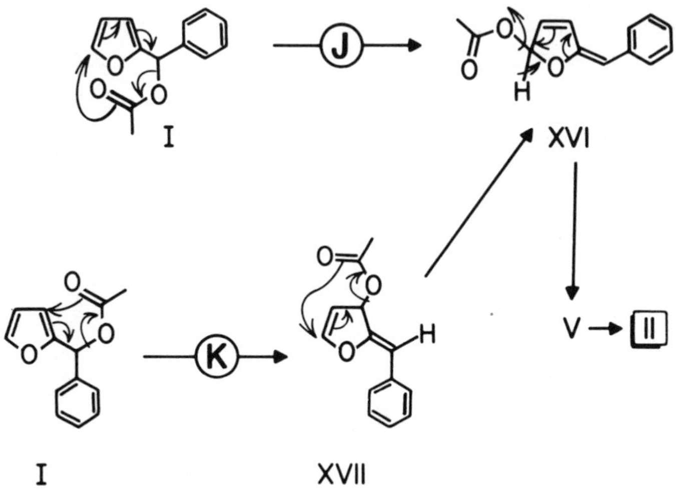

SCHEME 54.6

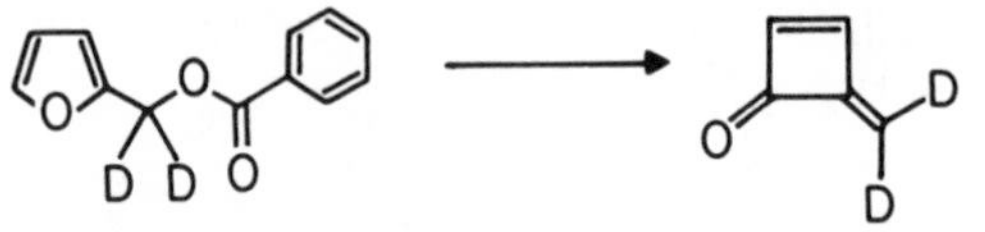

XVIII XIX

SCHEME 54.7

Route **K** of Scheme 54.6 describes this pathway as originally proposed by the authors of this work[1] (see also Problem 19).

Summing up, only routes **H, I,** and **K** are likely to convey the real situation. Fortunately, deuteration experiments have been performed and have shed some light on this mechanism. Compound **XVIII** was allowed to rearrange upon heating[1] and a 40% yield of cyclobutenone derivative **XIX** containing more than 96% deuterium at the terminal methylene was recovered (see Scheme 54.7). Obviously, any mechanism that requires the intermediacy of an alkyne of type **III** cannot fulfill this requirement. However, the carbene route cannot be ruled out because the pyrolysis of the corresponding deuterated compound **I** yields a mixture of products containing 54% deuterium only. Although this certainly indicates that "the migration process (route **K**) accounted for a major share of **II** produced, the carbene route (route **I**) was still operative".[1,7] Finally, the pericyclic reaction represented by route **H** could have been taken seriously if it did not involve an unfavorable Huckel-type, eight-electron transition state.

As for compound **IV**, it might be derived from cummuleno aldehyde **XX**, which in turn would be produced from the first acetate migration that gave **XVII** (Step **K** of Scheme 54.6). Thus, the presence of **IV** provides additional support to the double acetate shift mechanism (see Scheme 54.8).

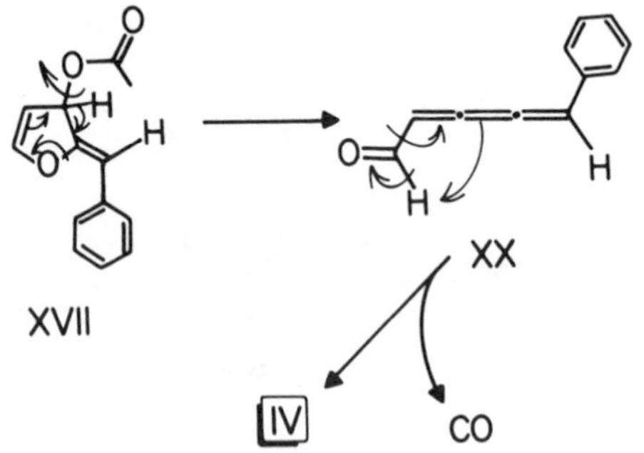

XVII XX IV CO

SCHEME 54.8

REFERENCES

2. Indeed, of the few known ketones of this type, only isopropylidene cyclobutenone is known besides **II.** See R. C. DeSelms and F. Delay, *J. Am. Chem. Soc.*, **95,** 274 (1973).
3. G. Chuchani, I. Martin, and M. E. Alonso, *Intl. J. Chem. Kinet.*, **9,** 819 (1977).
4. For the gas phase chemistry of carbenes see M. Jones, Jr., *Acc. Chem. Res.*, **7,** 415 (1974).
5. The 1,1 elimination as a source of carbenes is a well established process. See, among others, R. E. Lehr and J. M. Wilson, *J. Chem. Soc. Chem. Commun.*, 666 (1971); L. T. Scott and W. T. Cotton, *J. Am. Chem. Soc.*, **95,** 2708, 5416 (1973).
6. R. V. Hoffman and H. Schechter, *J. Am. Chem. Soc.*, **93,** 5940 (1971).
7. ... unless the 1,5-deuterium transfer is involved in the isomerization of **V** into **III** (which is a full 50% of the isolated mixture and is thus in large supply) does not occur in a concerted fashion as shown in the reverse step C_{-1} of Scheme 54.2. Also, the occurrence of a free radical mechanism, not completely unexpected for a cummulene at high temperature, would certainly interfere with the deuterium balance by transferring hydrogens from other carbons.

PROBLEM 55

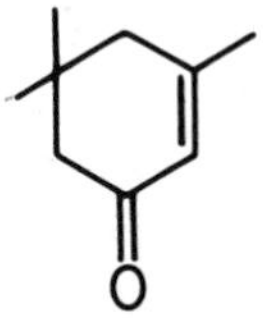

NaOH, H_2O

150°, 20 hr

(14% YIELD)

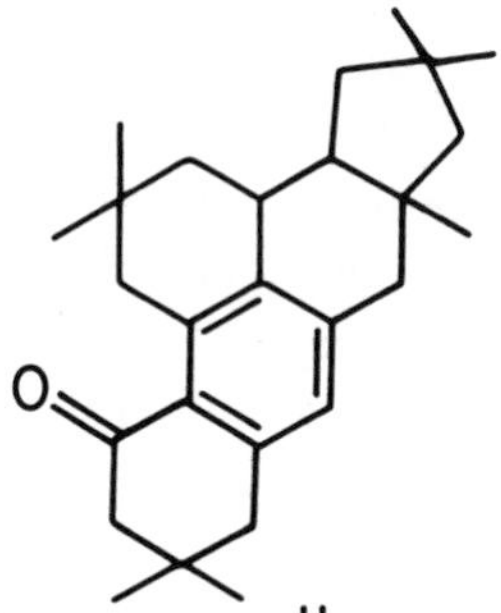

I

II

OR

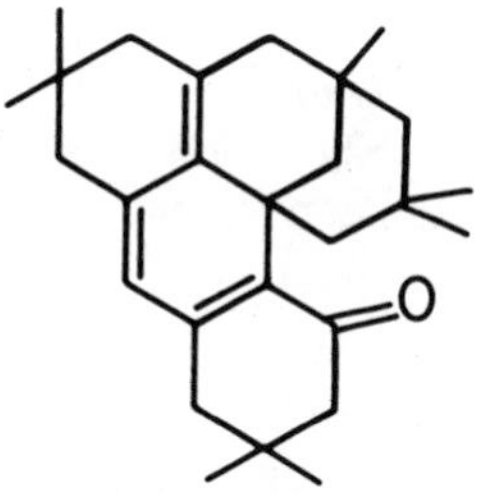

III

1. J. A. Bertrand, D. Cheung, A. D. Hammerich, H. O. House, W. T. Reichle, D. Vanderveer, and E. J. Zaiko, *J. Org. Chem.*, **42,** 1600 (1977).

PROBLEM 55
A Case of X-Ray Data Yielding the Wrong Structure due to Interference of Organic Chemistry

The complex process of structure elucidation usually includes chemical degradation, spectroscopic methods, and correlation analysis with known, related compounds. Occasionally, however, the feasibility of a mechanism that accounts for the formation of the problem compound structure—if its precursor is structurally clear—can also be used as an additional criterion. During the course of a reinvestigation of the reaction described here, serious difficulties were found with structural assignments of the isolated compound. To a number of spectroscopic incompatibilities was added the authors inability to find reasonable mechanistic routes that could explain the generation of a structure suggested by earlier researchers.

The two postulated structures, the old and the new one, have been drawn without indicating which is the correct one, for us to discuss their feasibility of formation from **I** and base, and to make a choice on a mechanistic basis alone. It is fair to say that both structures were determined by correct interpretations of X-ray diffraction analysis, the final authority in structural determination methods today. Paradoxically however, one of the structures is definitely wrong!

Comparison of empirical formulas indicates that compounds **II** and **III** are isomeric, and correspond to an apparent trimer of isophorone **I** minus two molecules of water. This circumstance suggests that the carbonyl group of **I** is a center of C–C bond formation by way of aldol condensation followed by elimination of the resulting alcohol. Predictably, enone **I** should be prone to polymerize under alkaline conditions due to its two electrophilic carbons and three potential carbanionic centers (see Scheme 55.1).

This sole feature opens a considerable number of possible trimerization

BASE
Nu
A
B

SCHEME 55.1

pathways. In actual fact, three different dimers have been identified in the past (plus two more postulated here).[2] It is also important to recognize that **II** and **III** contain the isophorone system bonded at the C-3 methyl substituent, in addition to the obvious nuclear carbons. This will require the contribution of resonance form **B** in at least two of the three fragments.

Dissection of compound **II** (Scheme 55.2) readily reveals two enone components. The third, however, appears less visible. The dotted line in this fragment not only suggests the presence of an isophorone molecule that has undergone 1,4 addition of a **B** type enolate, but also indicates that the fragmentation of the α C–C bond of the carbonyl sector, must have occurred with concurrent bonding of the liberated methylene to the terminal methylene of **B,** namely C_a–C_b.

The last conclusion calls for rather bold mechanistic proposals. Of the various ones we were able to conceive (not shown), the less preposterous—shown in Scheme 55.3—begins with the simple coupling of enolate **B** with a neutral molecule of isophorone by way of 1,4 addition. Base treatment of dimer **IV** might lead to further intramolecular interaction to give at least three reasonable adducts **V, VII,** and **VIII.** Then the latter would undergo isomerization in two divergent ways, both entailing the ring disconnection suggested by fragmentation analysis of Scheme 55.2. The first of these routes, however, requires the 1,2 migration of a hydride in order to allow for the new C_a–C_b bond, as indicated in **VIII.**

Although molecular models showed appropriate sterochemistry and reasonable orbital overlaps, there appears to be no precedent in the literature of such pinacol-type rearrangements. At any rate, there is always a first time, and the resulting *cis*-ring fused system **IX** would be opened for double 1,2 addition at

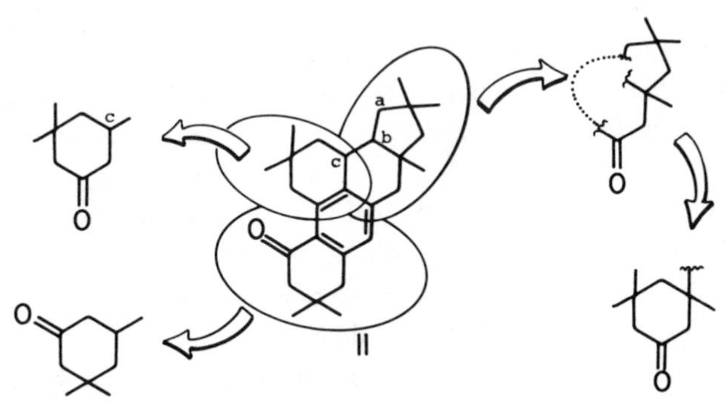

SCHEME 55.2

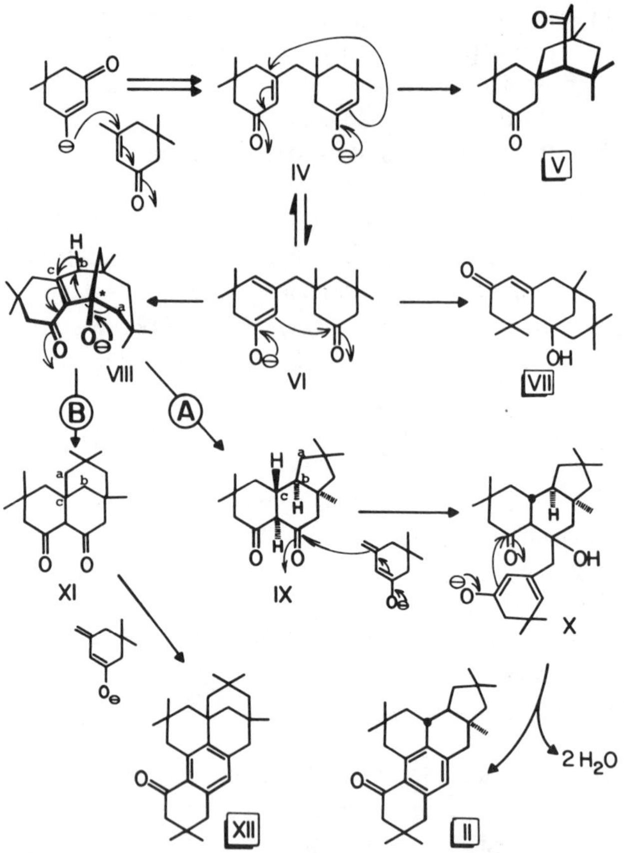

SCHEME 55.3

the carbonyl carbon by the third isophorone molecule followed by loss of 2 mol of water and aromatization, thus affording **II.**

The justification of compound **III** also requires unorthodox steps. While the bicyclic system in this compound is certainly an indication that **III** stems from the key intermediate **VIII,** it also implies that the tertiary C–O bond at the bridgehead carbon of **VIII** (denoted C*) must be somehow replaced by a C–C bond. On the one hand, however, C* is not amenable to a S_N2 displacement since the back entrance to the C–O bond is obstructed by the bulky rest of the molecule. On the other hand, although we have already seen cases in which bridgehead carbenium ions were invoked (see Problems 44 and 53) compound **VIII** as its alcohol or chloro derivatives is remarkably reluctant to yield a carbocation at this C* position, least of all in alkaline medium. How do we explain what happens?

Enter astonishing chemistry here. Not everything that elicits our sense of awe ought to be inextricably complicated. In fact, the principle that underlies the following arguments is utterly simple. Everyone knows that, along with nucleophilic substitutions, addition processes are helpful in building carbon–carbon bonds. With the exception of step **A** we have done all but that in making trimers **II** and **XII.** Therefore, the electrophilic character of C* might well be developed by placing on it a C=C bond by means of 1,4 elimination of water. This operation converts C* into the β carbon of an α,β-unsaturated ketone, thus making it suitable for a Michael 1,4 addition of enolate **A.** This makes **XIV** (see Scheme 55.4).

Yes, **XIV** contains a bridgehead olefin. However, anti-Bredt compounds are not anymore the indiscretions they used to be years ago. Although still strained, they are capable of independent existence at room temperature, so much so that several members of this odd family have been synthesized.[3] Then,

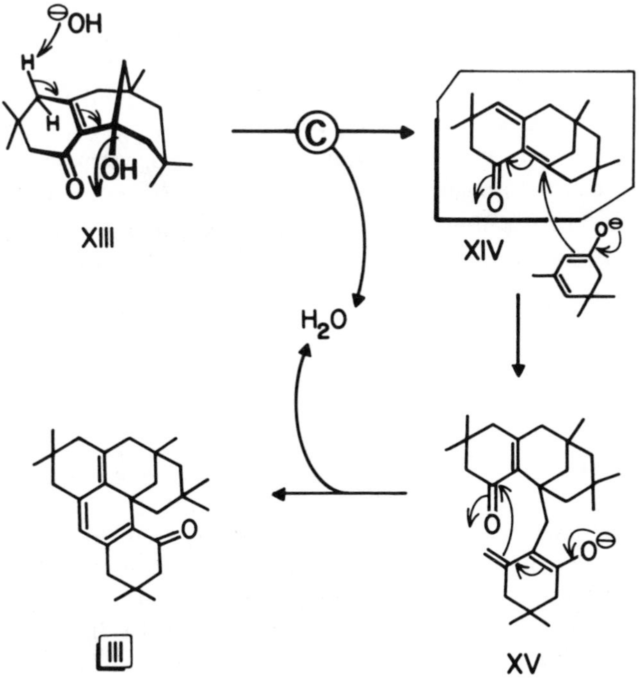

XIII XIV XV III C H_2O

SCHEME 55.4

anti-Bredt compound **XIV** would find its way to compound **III** by a second intramolecular 1,2 addition of a **B**-type enolate.

Now, the question arises—which is the correct structure, **II** or **III**? A final choice with 100% certainty between **II** and **III** cannot be made on the basis of a reaction mechanism alone, because both pathways go through rough terrain. However, the lack of precedent for step **A** of Scheme 55.3, the actual existence of bridgehead olefins, and a great deal of prejudice on our part, caused by our knowing beforehand that **III** is the correct structure, make us favor Scheme 55.4 as the more reasonable mechanistic interpretation.

The question that remains to be answered now is—where is the mistake in the structural interpretation of **II**? Earlier researchers obtained X-ray analysis of **II** in the days when heavy atoms had to be incorporated into the organic structures in order to obtain good diffraction patterns. Compound **II** was therefore not observed directly but as its iodo derivative. This halogen was placed at the aromatic unsubstituted carbon. It is likely that **III** was unknowingly isomerized into **II** by this halogenation procedure (iodine and silver trifluoroacetate). The reader is invited to justify this transformation and is left wondering why the ketone ring in **II** (which should not have undergone any transformation in the process) was detected in **II** in the opposite direction than that found in **III.**

Finally, it is important to emphasize that the value of a mechanism alone cannot be overestimated as a decisive criterion in structural elucidation. This is underscored by the overwhelming structural variation open to the combination of three molecules of isophorone (of which just a small fraction was described here). A word in favor of mechanisms: A logical reaction mechanism standing behind any given structure—if one knows its chemical origin—is, to say the least, reassuring.

REFERENCES

2. G. Buchi, J. H. Hansen, D. Knutson, and E. Koller, *J. Am. Chem. Soc.*, **93,** 5517 (1958).
3. For recent reviews see R. Keese, ''Methods of Preparation of Bridgehead Olefins,'' in *New Synthetic Methods*, Vol. IV, Verlag Chemie, Basel, pp. 95–130 (1979); K. H. J. Shea, *Tetrahedron*, **36,** 1683 (1980). A most extraordinary naturally occurring bridgehead olefin, taxol (**XVI**) is on record. See M. C. Wani, H. L. Taylor, M. E. Wall, P. Coggon, and A. T. McPhail, *J. Am. Chem. Soc.*, **93,** 2325 (1971). Even dienic bridgehead olefins have been recently reported. See Y. Tobe, T. Kishimura, K. Kakiuchi, and Y. Odaira, *J. Org. Chem.*, **48,** 551 (1983).

XVI

(TAXOL)

PROBLEM 56

H_3CO OCH_3 Cl Cl Cl Cl O CH_3

I

H_2SO_4, H_2O

0°, 15 min

O Cl Cl Cl $COOCH_3$ CH_3

II

1. R. R. Sauers, R. Bierenbaum, R. J. Johnson, J. A. Thich, J. Potenza, and H. J. Schugar, *J. Org. Chem.*, **41,** 2943 (1976).

PROBLEM 56
A Cyclopropyl Ketone from a Polychlorinated Acetal: A True Challenge

Polychlorinated hydrocarbons, besides being notoriously persistent pollutants, are exceedingly interesting compounds that frequently pose arduous mechanistic problems.[2] The deep seated transformation of **I** into **II** proves to be no exception. This challenging mechanistic fortress will force us to appeal to almost every available reasoning weapon to attempt its assault.

The cornerstone of our rationale is laid down by comparison of empirical formulas of starting material and product. This indicates that one carbon, three hydrogens, and a chlorine atom are removed in the transformation of the starting material. That the lost carbon does not belong to the molecular backbone is suggested by the fact that the number of C–C bonds—11 in all—remains unchanged. Therefore, the methyl carbon must be released as part of a methoxy group, since this is the only carbon-containing removable group in the molecule. In that case, the departed oxygen would have to be replaced by another oxygen atom from, say water, in order to maintain the atomic balance.

Furthermore, bond budgeting indicates that, if the missing C–Cl bond in **II** is not reflected in an increase of C–C bonds (*remember:* there are 11), or in the formation of a C=C unsaturation produced by elimination, the departure of the chloride ion must be translated necessarily into an effective increase in the oxidation level in going from **I** to **II.** In fact, we see that an ester is produced in the process while no group in **I** has a comparable oxidation level. All this means that we should use the ejection of chloride ion for the formation of the carboxylate unit of **II.** This in turn strongly implies that C* (see Scheme 56.1) is the most likely source of the ester carbon. Furthermore, C* is part of a potentially labile oxetane, whose separation gives ample opportunity for change.

A first consequence of this line of thought would be that either one of the C–C bonds stemming from C* should be broken at some point along the reaction course. Of these two, the disconnection of $\overset{\frown}{C^{*}\text{–}C_a}$ is perhaps the most favorable, owing to the ease of cleavage of a methoxy substituent in acidic media. All this leaves the acetal function as the most probable generator of the ketone group in **II.**

With these considerations in mind, a good starting point of the mechanistic sequence would be either the acid assisted ring opening of the oxetane function

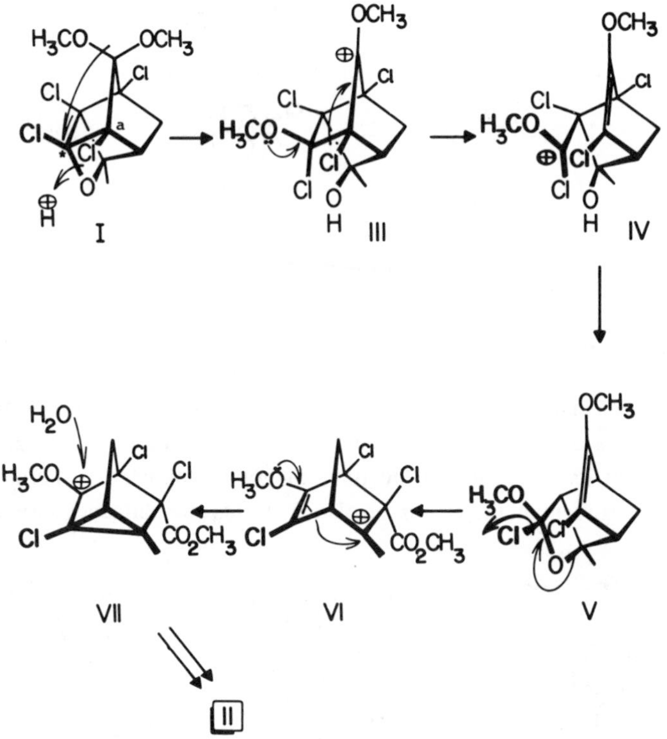

SCHEME 56.1

or the extrusion of methanol from the acetal group. The authors of the original article[1] preferred the first option (Scheme 56.1) and added the unusual, but still precedented[3] 1,3-shift of the endo-methoxy group. The aforementioned ester was then constructed from the resulting carbocation **III** by way of the indicated C–C bond separation, regeneration of the oxetane **V** and reopening of the same ring in a different direction to give cation **VI.** The trapping of this cation with the nucleophilic enol ether function and hydrolysis finished their sequence.

A much simpler scheme avoids the methoxy shift and the somewhat cumbersome opening–recombination–reopening of the oxetane ring. This results by following our first option, namely, begin with exclusion of methanol from the acetal. The resulting carbenium ion (**VIII**) would provide the driving force needed for the disconnection of the bicyclic carbocycle. We said before that this C–C bond cleavage was a requirement for the construction of the ester function of **III.**[4] Then, methanolysis of the acid chloride fragment that results

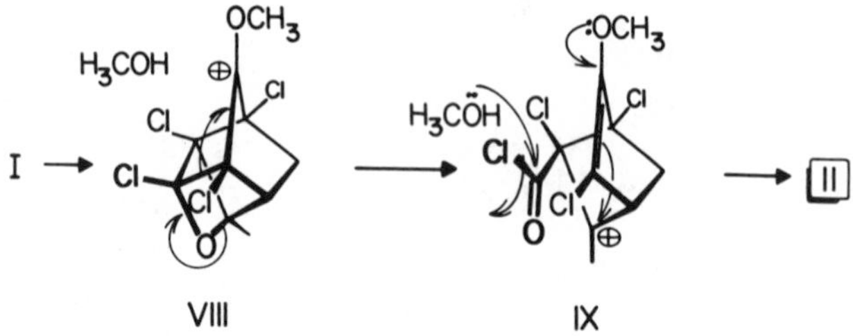

SCHEME 56.2

(**IX**) would be performed by a methanol molecule trapped in the solvent cage, much in the way postulated in Problem 36 (see Scheme 56.2).

The basic concepts advanced at the beginning of our discussion permit other possible routes of conversion. Scheme 56.3 depicts one of these hypotheses, which, for all its obeisance to the rules of good chemistry, is perhaps more a mental divertissement than anything else. Yet, this new sequence serves its pur-

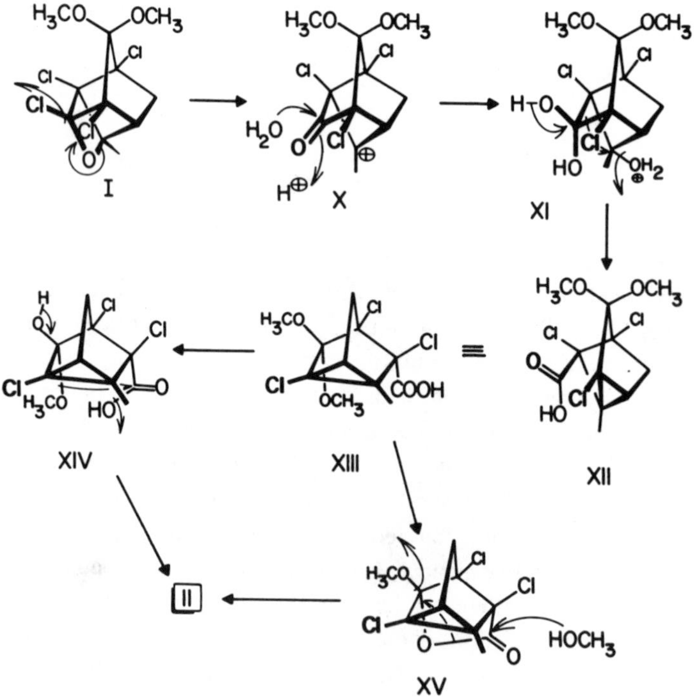

SCHEME 56.3

pose as an illustration of the fact that, even the most challenging and demanding problems do have more than one solution within the reach of the average intellect as long as some of the tremendous potential of people's wits is put to good use.

REFERENCES

2. See, for example, L. S. Besford, R. C. Cookson, and J. Cooper, *J. Chem. Soc.*, *C*, 1385 (1967); K. V. Scherer, Jr., R. S. Lunt, III, and G. A. Ungefug, *Tetrahedron Lett.*, 1199 (1965); K. V. Scherer, Jr., *Tetrahedron Lett.*, 2077 (1972); R. J. Stedman, and L. S. Miller, *J. Org. Chem.*, **32,** 35 (1968); W. G. Dauben and N. L. Reitman, *J. Org. Chem.*, **40,** 841 (1975).
3. P. G. Gassman, J. L. Marshall, and J. G. MacMillan, *J. Am. Chem. Soc.*, **95,** 6319 (1973).
4. The readers should convince themselves that breaking the other C*–C bond by way of a 1,3-shift towards the positive carbon leads to a completely different, though interesting set of structures. This will become a rewarding exercise.

AUTHOR INDEX

REACTION INDEX